SYNOPTIC ANALYSIS AND FORECASTING

SYNOPTIC ANALYSIS AND FORECASTING

An Introductory Toolkit

SHAWN MILRAD

ELSEVIER

Elsevier
Radarweg 29, PO Box 211, 1000 AE Amsterdam, Netherlands
The Boulevard, Langford Lane, Kidlington, Oxford OX5 1GB, United Kingdom
50 Hampshire Street, 5th Floor, Cambridge, MA 02139, United States

Notices

Knowledge and best practice in this field are constantly changing. As new research and experience broaden our understanding, changes in research methods, professional practices, or medical treatment may become necessary.

Practitioners and researchers must always rely on their own experience and knowledge in evaluating and using any information, methods, compounds, or experiments described herein. In using such information or methods they should be mindful of their own safety and the safety of others, including parties for whom they have a professional responsibility.

To the fullest extent of the law, neither the Publisher nor the authors, contributors, or editors, assume any liability for any injury and/or damage to persons or property as a matter of products liability, negligence or otherwise, or from any use or operation of any methods, products, instructions, or ideas contained in the material herein.

Library of Congress Cataloging-in-Publication Data
A catalog record for this book is available from the Library of Congress

British Library Cataloguing-in-Publication Data
A catalogue record for this book is available from the British Library

ISBN: 978-0-12-809247-7

For information on all Elsevier publications
visit our website at https://www.elsevier.com/books-and-journals

  Working together
to grow libraries in
developing countries

www.elsevier.com • www.bookaid.org

Publisher: Candice Janco
Acquisition Editor: Laura S Kelleher
Editorial Project Manager: Hilary Carr
Production Project Manager: Nilesh Kumar Shah
Cover Designer: Greg Harris

Typeset by SPi Global, India

CONTENTS

PREFACE

This text emanated from the course notes of my Introduction to Weather Analysis and Forecasting course at Embry-Riddle Aeronautical University, and a similar course that I taught at the University of Kansas in 2010–11. The assumption here is that students have already taken at least one course in introductory meteorology and have been exposed to qualitative concepts such as pressure, temperature, humidity, density, and wind.

The intended audiences of this text include meteorology students in their second or early third year of undergraduate studies, students minoring in meteorology, and qualified professionals in related fields (e.g., aviation, emergency management) aiming to gain a better knowledge of weather analysis and forecasting. Because of the wide range of expertise among the intended audiences, this text is almost entirely qualitative. It is not expected that readers will have much knowledge of the calculus-based governing equations of the atmosphere. For undergraduate meteorology or atmospheric science majors, this text is intended as a bridge between introductory courses and calculus-based dynamics and thermodynamics courses.

A wide range of topics are covered in this text, from basic terminology, surface station plots and METAR code, to analysis of physical mechanisms associated with vertical motion and the intensity and motion of synoptic-scale weather systems. At the conclusion of this text, the reader should have accumulated a solid knowledge base in weather analysis and basic forecasting, which can be applied in either more advanced calculus-based courses or weather-related careers and applications.

ACKNOWLEDGMENTS

This text would never have been possible without the science, education, and career knowledge gained under the tutelage of John Gyakum and Eyad Atallah of McGill University, and Mark Wysocki of Cornell University. Much of the foundation of my Introduction to Weather Analysis and Forecasting course was developed based on Mark Wysocki's undergraduate meteorology courses at Cornell University. Mark and Eyad are the two most gifted meteorology instructors I have ever known and I am forever grateful for the lessons that they taught me in science and in how to be a teacher.

My colleagues at Embry-Riddle Aeronautical University, specifically Tom Guinn and Debbie Schaum, have been invaluable resources of knowledge and support during my academic career. I am fortunate to have worked in such a collaborative environment where the faculty truly cares about their students and undergraduate education. And special thanks to all of my students, past and present, at McGill University, University of Kansas, and Embry-Riddle Aeronautical University, meteorology majors, minors, or those just with an interest in weather, for providing me with the inspiration to continue to try to pass along my weather analysis and forecasting knowledge as best I can.

I also could not have completed this endeavor without the assistance and support of my technical editor, longtime great friend, and fellow meteorologist Briana Gordon. She is the best wordsmith I have ever known and her combination of meteorological knowledge and command of the English language has been incredibly helpful. I am forever thankful for her patience, work ethic, and guidance during this process. To my other close friends and family, especially my mother Jan, brother Jared, and stepmother Gloria, your love and support never goes unnoticed. To my great friend Rachael, who has been selflessly supportive and encouraging of my endeavors. And of course to Jen, whose unwavering love and support has meant more to me than I can ever say or reciprocate.

Finally, to my late father, Martin Milrad, who never talked much about himself, but whose love and encouragement was always noticed. During my times of personal doubt and struggle, he always encouraged me to persevere and be the best person I could be. He is so missed, but I know he is smiling and cracking a joke at me every day from wherever he is.

1

METEOROLOGICAL CONVENTIONS

1.1   Time Zones and Zulu (Z) Time

Weather analysis and forecasting are global pursuits. The forecaster should not limit him or herself to learning only about their city, region, country, or continent. Because numerical weather prediction models and forecasts are produced across the globe, it is crucial that the meteorological enterprise operates largely independent of time zones. In fact, the word *synoptic* is derived from Greek, meaning "at the same time."

Meteorological observations and forecasts use Universal Coordinated Time (UTC), otherwise known as Zulu (Z)[1] or Greenwich Mean Time (GMT), which operates on a 24-h clock. This system was first devised in the United Kingdom and is based on the time at Greenwich, England, which is located on the Prime Meridian (0° longitude). In order to become familiar with typical meteorological analysis and forecast times, it is crucial to know how to convert from Z to your local time zone and vice versa.

The conversion from Z to your local time requires knowledge of how many time zones you are away from Greenwich, England, and what time of year it is (i.e., whether it is Daylight Saving Time). For example, 12Z is noon in Zulu time; at the same time, the US east coast is 7 a.m. local time in the winter (Eastern Standard Time) and 8 a.m. local time in the summer (Eastern Daylight Time). In California and British Columbia, 12Z is 4 a.m. Pacific Standard Time and 5 a.m. Pacific Daylight Time, respectively. Table 1.1 details the time conversion and local times for various locations across North America at key meteorological observation times.

---

[1] The term Zulu is simply from amateur radio or military conventions, where the word for the letter "Z" is Zulu.

Synoptic Analysis and Forecasting. https://doi.org/10.1016/B978-0-12-809247-7.00001-6

Table 1.1 Examples of Time Conversions From Zulu (Z) Time for Specific Locations Across North America

Time Zone	Cities Included	Hours Behind Z	00Z	12Z
Atlantic Standard Time (AST)	Halifax, Hamilton (Bermuda), San Juan (Puerto Rico)	4	8 p.m. AST	8 a.m. AST
Eastern Standard Time (EST)	New York City, Miami, Toronto	5	7 p.m. EST	7 a.m. EST
Central Standard Time (CST)	Chicago, Houston, Winnipeg	6	6 p.m. CST	6 a.m. CST
Mountain Standard Time (MST)	Denver, Albuquerque, Calgary	7	5 p.m. MST	5 a.m. MST
Pacific Standard Time (PST)	Los Angeles, Seattle, Vancouver	8	4 p.m. PST	4 a.m. PST
Alaska Standard Time (AKST)	Juneau, Anchorage, Nome	9	3 p.m. AKST	3 a.m. AKST
Hawaii Standard Time (HST)	Honolulu, Hilo, Pearl City	10	2 p.m. HST	2 a.m. HST

For Daylight Saving Time, one fewer hour should be subtracted from Z.

1.2 Common Units of Measurement

Most meteorological variables and calculations are handled using standard metric (SI) meters-kilograms-seconds (MKS) units, although there are some notable exceptions that we discuss in this section. There are also important regional differences with certain variables; while most countries use the metric system, the United States and some other countries still use the imperial system for specific meteorological variables.

1.2.1 Temperature

Although most meteorological calculations that involve temperature use the Kelvin (K) scale, K is not used much in operational analysis and forecasting. The overwhelming majority of the world uses degrees Celsius (°C) to report both surface and upper-air temperatures, even if certain countries still use the imperial system for other variables. The one major exception to this is surface temperature in the United States, which is still reported in degrees Fahrenheit (°F). Therefore it is useful for the forecaster to know how to convert between °C and °F, and vice versa.

The equation to convert from °C and °F is given in Eq. (1.1), while the inverse is shown in Eq. (1.2).

$$°C = 5/9(°F - 32) \qquad (1.1)$$

$$°F = 9/5°C + 32 \qquad (1.2)$$

It is also useful to memorize key threshold temperature values in °C and the corresponding values in °F. Table 1.2 shows some of these important thresholds, including the freezing point of water, 0°C (32°F). Note that the two temperature scales "converge" at −40, such that −40°C is equivalent to −40°F.

1.2.2 Wind Speed and Direction

Wind speeds are typically measured in knots (kt), where 1 kt is defined as 1 nautical mile per hour. Some countries use meters per second (m s^{-1}), miles per hour (mph), or kilometers per hour (kph)

Table 1.2 Key Values of Temperature in Degrees Celsius (°C) and Their Corresponding Values in Degrees Fahrenheit (°F)

°C	°F
−40	−40
−30	−22
−20	−4
−10	14
−5	23
0	32
5	41
10	50
15	59
20	68
25	77
30	86
40	104
50	122

instead of kt. Standard meteorological observations are most commonly either in kt or m s^{-1}, so it is useful to know that 1 m s^{-1} = 1.94 kt, and to be able to convert back and forth between the two.

Meteorological wind directions are reported in degrees, using the direction the wind is *coming from*. This convention can require an adjustment for novice observers and forecasters who may be used to other coordinate systems, such as the one used in mathematics. Fig. 1.1 shows a graph of the meteorological wind direction coordinate system, where 360° is a wind *coming from* the north (northerly), 90° is an easterly wind, 180° is a southerly wind, and 270° is a westerly wind.

1.2.3 Pressure

In most countries, meteorological charts use the units of millibars (mb) or hectopascals (hPa) for atmospheric pressure, where 1 mb = 1 hPa. The mb can be converted to the MKS unit Pascal (Pa) by multiplying by 100, where 1 mb = 1 hPa = 100 Pa. Some aviation interests, particularly in the United States, use inches of mercury (inHg) for altimeter settings and readings. Therefore it is useful for aviation-oriented users and forecasters to know the conversion between mb and inHg, where 1 inHg = 33.865 mb. Using this conversion, the standard atmospheric sea-level pressure of 29.92 inHg is equivalent to 1013.2 mb (hPa). For the remainder of this text, we will generally use hPa for pressure, with the understanding that mb and hPa are interchangeable.

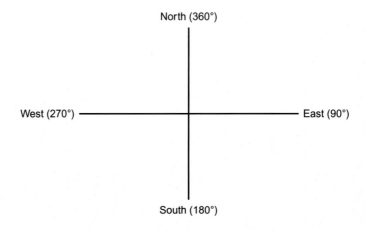

Fig. 1.1 Meteorological wind direction coordinate system, with the primary wind directions listed in degrees. The wind is always reported as the direction it is coming from, e.g., a north (northerly) wind is coming from the north.

1.3 Questions and Practice Exercises

1. It is 1945Z on January 25th of some year. What time is it in each of the following locations? Please give your answers in local time appropriate for the time of year.
   * Boston, Massachusetts.
   * Brandon, Manitoba.
   * Lubbock, Texas.
   * Kelowna, British Columbia.
   * Honolulu, Hawaii.
2. It is 0820Z on August 12th of some year. What time is it in each of the following locations? Please give your answers in local time appropriate for the time of year.
   * Halifax, Nova Scotia.
   * St. Louis, Missouri.
   * Edmonton, Alberta.
   * Eugene, Oregon.
   * Fairbanks, Alaska.
3. You measure the 12Z temperature in your favorite Canadian city to be 11.1°C. What temperature in this in °F?
4. You measure the 00Z temperature in your favorite US city to be 24°F. What temperature in this in °C?
5. Give the precise wind direction in degrees for each of the following descriptors:
   * Southwesterly.
   * Northeasterly.
   * Northwesterly.
6. A wind coming from 120° indicates that the wind is coming from roughly which direction?
7. You measure the wind speed at your location to be 15 kt and the atmospheric pressure to be 998.7 mb. An aviation user requests that you convert wind speed to m s^{-1} and pressure to inHg. What are your measured values in those units?

2

SURFACE OBSERVATIONS AND INSTRUMENTATION

2.1 Automated Surface Observing Systems

2.1.1 What is an ASOS?

Most major airports still employ human weather observers, some of whom work double duty as air traffic controllers. However, in many countries, larger airports rely partly or entirely on automated systems to take regular surface weather observations. Major airports in the United States and Canada use a version of the Automated Surface Observation System (ASOS).[1] In the United States, most large- and medium-sized (>900 total) airports use ASOS, while smaller airports use Automated Weather Observing Systems (AWOS), which typically have fewer instruments and thus report fewer meteorological variables. There are several varieties of AWOS, with differing levels of measurement capabilities.[2] In some countries, major airports have automated systems that, for all intents and purposes, are equivalent to ASOS, but are referred to as AWOS. To minimize confusion here, we will use the term ASOS to describe the standard instrumentation installed at most medium-sized or larger commercial airports.

---

[1] This is a term used exclusively for the machines developed and maintained by the National Weather Service (NWS) and Federal Aviation Administration (FAA) for use at major US airports. While equivalent equipment is found in other countries, it is not technically referred to as ASOS. Further information can be found online at: http://www.nws.noaa.gov/asos/.

[2] The FAA has an online database of ASOS and AWOS stations across the United States, searchable by airport code or state, available online at: https://www.faa.gov/air_traffic/weather/asos/.

Synoptic Analysis and Forecasting. https://doi.org/10.1016/B978-0-12-809247-7.00002-8

2.1.2 ASOS Instrumentation

Fig. 2.1 shows an example of an ASOS station with the standard instrument suite. On the left is a 10-m high tower that is used to measure wind speed and direction; this is the only instrument in the ASOS that is higher than 2 m. Wind is measured at a higher altitude to minimize the impacts of turbulent eddies close to the surface. At the top of the tower are two instruments: a wind vane to measure wind direction, and a sonic anemometer that measures wind speed. While the wind vane is one of the oldest weather instruments, the sonic anemometer is one of the newest, with many ASOS stations having only recently been upgraded.

Wind direction is defined as the direction that the wind vane is pointing *toward*, or the direction the actual wind is coming *from*, and is reported to the nearest 10° (see Chapter 3). The sonic anemometer determines wind speed by sending sound pulses back and forth between its sensors and measuring the speed of sound as it is changed slightly by the wind. The sonic anemometer recently replaced the three-cup anemometer at ASOS stations. In addition to being able to measure both faster (>125 kt) and

Fig. 2.1 Example of an ASOS station with the standard equipment. From left: wind tower and sonic anemometer, tipping bucket rain gauge, hygrothermometer, present weather sensor, central processing unit (CPU) and pressure sensors, laser beam ceilometer, freezing rain sensor, and visibility sensor. Image courtesy of the COMET website at: http://meted.ucar.edu/ of the University Corporation for Atmospheric Research (UCAR), sponsored in part through cooperative agreement(s) with the National Oceanic and Atmospheric Administration (NOAA), US Department of Commerce (DOC). ©1997–2016 University Corporation for Atmospheric Research. All Rights Reserved, and available online at: http://www.meted.ucar.edu/emgmt/hazwx/media_gallery.php.

slower (<2 kt) wind speeds than the old three-cup anemometer was capable of, the sonic anemometer is more reliable in winter. The sensors on the sonic anemometer are capable of applying heat to melt ice and snow, whereas the three-cup anemometer would frequently be weighed down and/or frozen by ice and snow. Finally, the sonic anemometer is also able to take more frequent measurements; currently, a 3-s average speed is used for gusts and peak winds (see Chapter 3), while the older anemometers used 5-s averages.

The second instrument from left in Fig. 2.1 is the tipping bucket rain gauge, which is used to measure precipitation accumulation. Fig. 2.2 shows a closer view of the tipping bucket, which has a wind shield around it to limit the amount of precipitation that can escape in windy conditions. When 0.01 in. of liquid-equivalent precipitation fall into the bucket, it tips to one side, sending an electronic signal to the ASOS CPU. The bucket

Fig. 2.2 Example of a tipping bucket rain gauge, with the heating elements located at the bottom. From the National Weather Service (NWS) ASOS User's Guide, available online at: http://www.nws.noaa.gov/asos/pdfs/aum-toc.pdf.

subsequently tips for each additional 0.01 in. of precipitation. It is important to remember that the tipping bucket only measures liquid-equivalent precipitation; if frozen precipitation (e.g., ice pellets, snow) falls into the bucket, the heating elements at the bottom (Fig. 2.2) melt it into liquid. However, heating frozen precipitation also results in some sublimation and evaporation; as a result, frozen precipitation amounts are often underreported by the ASOS, making it vital that major airports employ a human weather observer to accurately measure snow and ice.

The third instrument from left in Fig. 2.1 is the hygrothermometer, which measures temperature and dew point. A photo of the hygrothermometer is shown in Fig. 2.3. This mushroom-shaped instrument uses a platinum wire Resistive Temperature Device (RTD) to measure both temperature and dew point. The RTD works on the principle that electrical resistance is dependent on temperature. Older (AWOS) equipment may still use the "chilled mirror" device to measure dew point, but it was replaced by the RTD in ASOS equipment due to persistent maintenance issues.

Next from the left in Fig. 2.1 is the present weather sensor, also shown in Fig. 2.4. This instrument is also known as the Light Emitting Diode Weather Identifier (LEDWI) and uses an infrared beam to detect precipitation type and intensity. The infrared beam is sent between the two camera-shaped objects (a transmitter and

Fig. 2.3 Example of the ASOS hygrothermometer, with the RTD contained in the mushroom-shaped instrument. From the NWS ASOS User's Guide, available online at: http://www.nws.noaa.gov/asos/pdfs/aum-toc.pdf.

Fig. 2.4 Example of the ASOS present weather sensor (LEDWI). From the NWS ASOS User's Guide, available online at: http://www.nws.noaa.gov/asos/pdfs/aum-toc.pdf.

a receiver), and the LEDWI measures the scattering of the infrared beam, which allows the instrument to assess precipitation type and intensity. The LEDWI can struggle in mixed or frozen precipitation situations, and will sometimes report unknown precipitation (UP, see Chapter 3) in those situations. This is another justification for employing human weather observers to augment mixed or frozen precipitation reports at major airports.

The large rectangular box in the middle of Fig. 2.1 is the CPU, which houses three barometers, the ASOS computer and power supply, and communication equipment. A closer image of the CPU is shown in Fig. 2.5. The three barometers in the lower half of the CPU are used to report station pressure, which is then

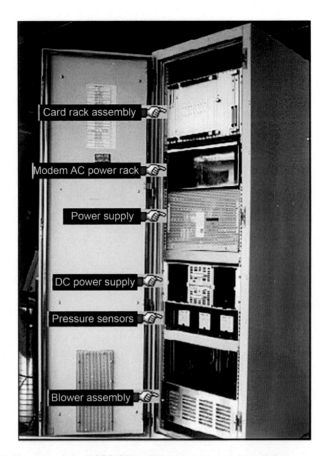

Fig. 2.5 Example of the ASOS CPU with the power supply and three pressure sensors (barometers), labeled. From the NWS ASOS User's Guide, available online at: http://www.nws.noaa.gov/asos/pdfs/aum-toc.pdf.

converted into altimeter setting and sea-level pressure (SLP). In order to get the most accurate pressure reading possible, the three barometer readings are averaged together. The pressure sensor is the most reliable in the entire ASOS, as it has an accuracy within 0.02%. The CPU and power supply are protected from elements and animals by a closed door. This minimizes malfunctions and ensures that the barometers, which are vital to aviation operations, continue to function even in harsh conditions.

Third from the right in Fig. 2.1 and shown in Fig. 2.6 is the laser ceilometer, which is used to determine cloud base heights and sky conditions. The ceilometer shoots an infrared laser beam through the top and calculates cloud height by timing the interval between the transmission and reception of the infrared pulse. Due to atmospheric light scattering, the ASOS ceilometer is only able to detect cloud bases up to 12,000 ft. Thus any clouds above 12,000 ft. must

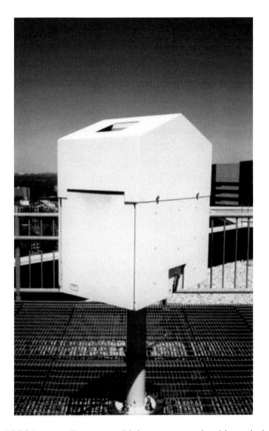

Fig. 2.6 The ASOS laser ceilometer, which measures cloud base heights and sky conditions. From the NWS ASOS User's Guide, available online at: http://www.nws.noaa. gov/asos/pdfs/aum-toc.pdf.

be reported by a human weather observer or they will not appear in the surface observation. Sky conditions are computed from a time history of clouds drifting over the ceilometer. The instrument is considered to be accurate within $1/8$ of cloud fraction and 100 ft. in cloud base height.

Beyond cloud base height, there are a few substantial limitations to the ceilometer. First, it can only see clouds directly above it, so it uses an automated algorithm to assess sky conditions away from the sensor, which can introduce error. Second, the ceilometer can only report a maximum of three different cloud heights. In addition, the ASOS assumes that all clouds are random and move with the wind, which is not always valid. For example, if there are cumulus clouds and light winds present at altitude, a patch of cumulus may remain above the sensor, but the ASOS algorithm will assume that the clouds have moved away and therefore report clear skies. Finally, falling precipitation can increase light scattering and thus degrade the accuracy of cloud height measurements.

The freezing rain sensor is the small instrument that is second from the right in Fig. 2.1, and shown in Fig. 2.7. It consists of a vibrating probe that detects the presence of ice by vibrating more slowly with increasing accumulation. The sensor measures ice every 15 min and reports accumulations to the CPU. Once ice is reported, the freezing rain sensor is heated to remove the ice, and the vibration cycle starts over again.

Last, but not least, is the visibility sensor, shown on the far right in Fig. 2.1 and also in Fig. 2.8. Similar to the LEDWI, the visibility sensor consists of a transmitter and a receiver. The transmitter emits a xenon light beam and the sensor computes visibility by measuring the amount of light that reaches the receiver. Different algorithms are used for daytime and nocturnal calculations. The ASOS visibility algorithm is also deliberately designed to respond quickly to decreases in visibility and respond slowly to increases in visibility, in order to maximize caution. The primary limitation to the visibility sensor is that it only sees a 0.75 cubic ft. volume of air

Fig. 2.7 The ASOS freezing rain sensor, with the vibrating probe shown at the top. From the NWS ASOS User's Guide, available online at: http://www.nws.noaa.gov/asos/pdfs/aum-toc.pdf.

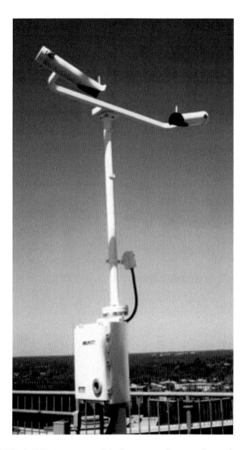

Fig. 2.8 The ASOS visibility sensor, with the transmitter and receiver both pictured. From the NWS ASOS User's Guide, available online at: http://www.nws.noaa.gov/asos/pdfs/ aum-toc.pdf.

between the transmitter and the receiver. Thus a particular sector of the airfield or station can have a smaller visibility due to patches of fog or other obscurations, which may not be reported properly by the ASOS. For situations such as reduced sector visibility (see Chapter 3), stations with manual weather observers are able to augment ASOS visibility reports, and thus provide a greater degree of precision.

2.2 Surface Station Plots: Conventions

2.2.1 Introduction to Surface Station Plots

The surface station plot is a useful tool for any forecaster. It provides a graphical depiction of surface conditions that is not in coded (e.g., METAR, see Chapter 3) or text form. The standard

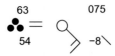

Fig. 2.9 Example of a surface station plot. Counterclockwise from top left: temperature, present weather, dew point, 3-h sea-level pressure (SLP) tendency, and SLP. Cloud cover and wind information are plotted at the center of the plot. Temperature and dew point are plotted in °F in the United States and in °C in most other countries.

appearance of the surface station plot is shown in Fig. 2.9. At the top left, temperature is plotted; in Fig. 2.9, the temperature is 63°F. Dew point is always plotted at the bottom left and is 54° F in Fig. 2.9. Note that the convention for surface station plots is to use °F in the United States and in °C in most other locations. Plotted on the left side in between the temperature and dew point is the present weather, which consists of any precipitation (e.g., rain, snow) and/or obscurations to visibility (e.g., fog, mist). The sample plot in Fig. 2.9 shows moderate rain (represented by three dots) with mist (represented by the two horizontal lines). We will discuss these symbols in more depth in Section 2.2.3.

Wind and sky cover information are located near the center of the surface station plot. In Fig. 2.9, the open circle at the center represents clear skies, while the wind stem and flags indicate that the wind is coming from the southeast at 15 kt. These parameters are discussed further in Sections 2.2.4 and 2.2.5. Finally, SLP information is placed on the right-hand side of the surface station plot. Plotted at the top right is SLP in tenths of a millibar, and plotted at the bottom right is the 3-h SLP tendency, which features a numeric value and a descriptive symbol. These are discussed in more detail in Section 2.2.6.

2.2.2   Temperature and Dew Point

Temperature and dew point are always plotted to the nearest whole degree. In the United States, and on any surface chart made in the United States, the convention for surface plots is to use °F for both variables. In most other countries, °C is used for both temperature and dew point. Particularly in the United States, it is important that the analyst or forecaster is able to convert between temperature units, because as we will learn in Chapter 3, METAR code always uses °C. For the surface station plot, negative temperatures and dew points are reported by putting a negative sign in front of the number. Finally, it is useful to remember that the dew point can be equal to, but never greater than the temperature.

2.2.3 Present Weather

The present weather section of the surface station plot is used to report precipitation (e.g., rain, snow), and obscurations to visibility (e.g., fog, mist). In all, there are 100 different present weather symbols as defined by the World Meteorological Organization (see online at: http://oceanservice.noaa.gov/education/yos/resource/JetStream/synoptic/ww_symbols.htm). Here, we will focus on the 21 most commonly used present weather symbols, which are shown in Fig. 2.10. There are three different intensity thresholds for rain and snow (light, moderate, heavy). Light rain (snow) is plotted using two dots (snowflakes), moderate rain (snow) using three dots (snowflakes), and heavy rain (snow) using four dots (snowflakes). It is a common mistake for the novice analyst or forecaster to use one, two, and three dots (snowflakes) for light, moderate, and heavy precipitation, respectively. Fig. 2.10

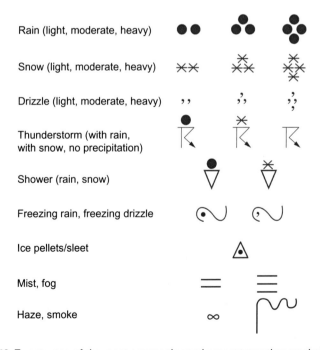

Fig. 2.10 Twenty-one of the most commonly used present weather symbols. For a full list of the 100 present weather symbols, see online at: http://oceanservice. noaa.gov/education/yos/resource/JetStream/synoptic/ww_symbols.htm.

also shows that drizzle is reported with the same convention as rain and snow, but with a comma-shaped symbol.

Occasionally, descriptors (see Chapter 3) such as showers and thunderstorms will be simultaneously occurring with precipitation. Descriptor symbols (Fig. 2.10) are plotted in the present weather section with one dot (for rain) or snowflake (for snow) plotted on top. In the case of a descriptor occurring with precipitation, the intensity of the precipitation will be reported in the METAR code but not plotted in the surface station plot. For example, heavy rain with a thunderstorm is plotted with the same symbol as light rain with a thunderstorm, with one dot on top of the thunderstorm symbol (Fig. 2.10).

The primary obscurations to visibility are fog, mist, haze, and smoke (Fig. 2.10). In a surface station plot, these can be plotted next to a precipitation (or descriptor) symbol if both are occurring. Since fog and mist are physically the same phenomena, with fog causing more severe reductions to visibility (see Chapter 3), they both use horizontal lines as present weather symbols. The only difference is fog is represented by three horizontal lines, whereas mist is indicated with two lines (Fig. 2.10). Haze and smoke are plotted using the infinity symbol and a plume-looking symbol, respectively. Unlike with descriptors, it is possible to plot precipitation intensity and an obscuration together. For example, in Fig. 2.9, moderate rain and mist are reported; note that the precipitation (or descriptor) is always plotted first, followed by the obscuration.

2.2.4 Wind Speed and Direction

Wind speed and direction information are found near the center of the surface station plot. The wind information consists of a long line, or wind stem, that indicates wind direction, and three possible symbols (flag, barb, and half-barb) to depict wind speed (Fig. 2.10). The wind stem points outward from the center of the station plot in the direction the wind is *coming from*. For example, in Fig. 2.9, the stem is pointing toward the southeast, indicating that the wind is coming from the southeast (southeasterly). In Fig. 2.11, the wind stem is pointing toward the west, indicating wind coming from the west (westerly).

Wind speed in a surface station plot is always reported to the nearest 5 kt. For example, if an ASOS is reporting 13 kt, the surface station plot would show 15 kt. Speed is expressed using three symbols, seen in Fig. 2.11: the flag represents 50 kt, the barb represents 10 kt, and the half-barb represents 5 kt. In Fig. 2.11, the wind

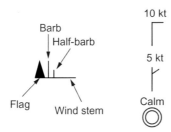

Fig. 2.11 Example of wind speed and direction information on a surface station plot. The wind stem points toward the direction the wind is coming from, the flag represents 50 kt, the barb represents 10 kt, and the half-barb represents 5 kt.

speed is being reported as 65 kt. In the Northern Hemisphere, the flags and barbs are always placed on the right-hand side of the wind stem looking outward from the center of the station plot; in the Southern Hemisphere, they are plotted on the left-hand side of the wind stem. To further differentiate between winds of 10 and 5 kt, a 5-kt wind is plotted using a half-barb slightly offset from the top of the wind stem (Fig. 2.11). A calm wind is indicated by an extra circle around the sky cover circle and no wind stem. Finally, some surface station plots will report wind gust information (if present) by writing "G" and the gust speed in kt on top of the flags or barbs.

2.2.5   Sky Cover

Sky cover is reported using the circle at the center of the surface station plot. Regardless of cloud height, the surface station plot always reports the maximum amount of sky coverage. For example, if skies are scattered (clouds covering approximately half of the sky) at 2500 ft., but overcast at 4000 ft., overcast is reported in the station plot. All possible symbols for sky cover are shown in Fig. 2.12. "Sky obscured," which is denoted by an "x" through the circle, refers to an indefinite ceiling observed from the ground, most commonly when the station is located in a fog bank. Sky cover in station plots will be related to the corresponding METAR code in Chapter 3.

2.2.6   Sea-Level Pressure (SLP)

The right-hand side of the surface station plot details SLP information. At the top right is SLP reported to the nearest tenth of a hPa. The leading "9" or "10" is always left off of the station plot. For

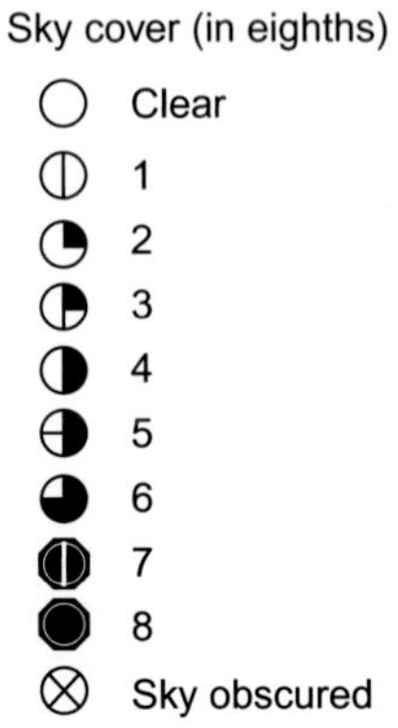

Sky cover (in eighths)

○	Clear
◍	1
◗	2
◑	3
◐	4
◑	5
●	6
◐	7
●	8
⊗	Sky obscured

Fig. 2.12 The nine possible choices for sky cover in a surface station plot, listed in eighths of cloud coverage.

example, in Fig. 2.9, 075 indicates that the current SLP is 1007.5 hPa. We know to add a leading 10 and not a 9 because 907.5 hPa is a value that would only be found at the center of a strong tropical cyclone. A good rule of thumb is to add a leading 9 when the first digit of the SLP is ≥ 5, and to add a leading 10 if the first digit is < 5. There are of course exceptions; for example, strong oceanic cyclones can reach 920–940 hPa in certain cases, while strong arctic high pressure systems can sometimes exceed 1050 hPa. In these situations, it is vital that the analyst or forecaster use his/her awareness of the current weather pattern in order to know which leading number to use.

At the bottom right of the plot is the 3-h SLP tendency. Because this is calculated every 3 h, it will only appear at 00, 03, 06, 09, 12, 15, 18, and 21Z. The SLP tendency is reported using a number and a symbol. The number is the 3-h SLP change to the nearest tenth of a hPa. The convention is to use a negative sign for negative SLP tendencies but no sign for positive SLP tendencies. In Fig. 2.9, the SLP has decreased 0.8 hPa in the past 3 h. The decimal point is always implied, but never written on the station plot. For example, a 1.8-hPa decrease would be encoded as −18.

To the right of the SLP tendency value is a symbol that describes how SLP has changed over the previous 3 h. Fig. 2.13 shows the nine possibilities for this symbol. When drawing a surface station plot, it crucial that the analyst or forecaster examines not only the magnitude of the SLP change but how it changed. In Fig. 2.9, SLP was continuously falling throughout the prior 3 h.

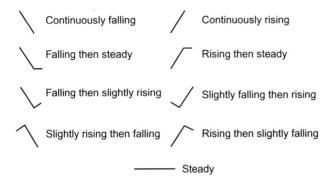

Fig. 2.13 The nine possible choices for the SLP tendency symbol.

2.3   Surface Station Plots: Examples

In Chapter 3, we will discuss and how surface station plots and METAR code are related. Prior to that, however, the forecaster must be able to independently create surface station plots when given data. Fig. 2.14 shows three examples of surface station plots from three very different weather situations. All temperatures and dew points are plotted in °F but can easily be converted to °C using the formula in Chapter 1. Fig. 2.14A shows a surface station plot from Newark, New Jersey, where heavy rain (four dots) and fog (three lines) are reported, and winds are from the northeast at 20 kt. In this case, a leading 9 is added to the SLP, indicating it is 997.6 hPa. SLP has fallen 1.6 hPa over the past 3 h, but steadied toward the end of the 3-h period.

Fig. 2.14B shows a station plot from a warm and mostly sunny day in Tampa, Florida, with few clouds and winds out of the southwest at 10 kt. Haze is reported and a leading 10 should be added to the SLP, making it 1021.2 hPa. The SLP tendency is weakly positive at 0.4 hPa over the past 3 h. Finally, Fig. 2.14C displays an example of a winter precipitation situation at Sioux Falls, South Dakota, with light snow (two snowflakes) and mist (two lines) being reported. The temperature and dew point are below freezing,

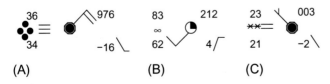

(A)   (B)   (C)

Fig. 2.14 Examples of surface station plots from (A) Newark, New Jersey, (B) Tampa, Florida, and (C) Sioux Falls, South Dakota, using standard US plotting conventions.

and the wind is from the northwest at 5 kt. Note that the 5 kt half-barb is not placed at the tip of the wind stem, so as to differentiate from the 10 kt barb. A leading 10 is added to the SLP, indicating it is 1000.3 hPa. SLP has decreased steadily but only slightly, 0.2 hPa over the past 3 h.

2.4 Questions and Practice Exercises

1. Explain how the tipping bucket measures accumulated precipitation, and how it measures frozen precipitation.
2. Describe how the laser ceilometer in the ASOS identifies cloud base heights and sky coverage, and list three limitations of this sensor.
3. Explain why the barometers in an ASOS are considered to be the most reliable of all the instruments.
4. Your favorite weather station is reporting present weather of heavy rain with a thunderstorm and mist. Draw the present weather section of a surface station plot using the appropriate symbols.
5. The following conditions are reported on a typical January day in Seattle, Washington. Draw a surface station plot using standard US plotting conventions.
   - Temperature: 43°F.
   - Dew point: 38°F.
   - Wind: 17 kt from the southwest.
   - Overcast skies.
   - Light rain and fog.
   - SLP: 999.2 hPa.
   - SLP 3 h ago: 1002.4 hPa.
   - SLP 2 h ago: 1000.8 hPa.
   - SLP 1 h ago: 999.3 hPa.
6. The following conditions are reported on a typical April day in Saskatoon, Saskatchewan. Draw a surface station plot using standard Canadian plotting conventions.
   - Temperature: 11°C.
   - Dew point: 4°C.
   - Wind: 23 kt from the northwest, gusting to 32 kt.
   - Scattered clouds.
   - Haze.
   - SLP: 1028.1 hPa.
   - SLP 3 h ago: 1025.8 hPa.
   - SLP 2 h ago: 1026.7 hPa.
   - SLP 1 h ago: 1027.3 hPa.

3

METAR CODE

METAR is a French acronym for "Météorologique Aviation Régulière" used to describe the code that is the standard surface observation reporting mechanism across the globe. METAR code was originally introduced for aviation use in the 1960s and was formally adopted as the global surface observation reporting standard in the late 1980s. Before METAR, some parts of the world used *synoptic code*, which is still utilized today for many ship reports.

Reporting all surface observations in METAR code is extremely useful for meteorologists, emergency managers, and anyone in the aviation sector, among others. To the first order, it is a relatively simple "language" to understand and decode, and having a global standard is a major benefit to all users. Although METAR code has too many rules and exceptions to cover in a single textbook chapter, by the end of this chapter the reader should feel comfortable reading and writing METAR code.

3.1 METAR Code: Conventions

Surface observation reports are required to be issued at least once an hour, either by a human observer or Automated Surface Observing System (ASOS, Chapter 2). A routine hourly report is called a METAR. In addition, there can be a maximum of 12 reports issued between routine hourly METARs; each one of these is called a special report (SPECI). In Fig. 3.1, notice the "METAR" at the beginning of the coded report; this indicates it is a routine hourly observation. Some sources (i.e., websites) may not specify METAR or SPECI at the beginning of the report. However, the user should be able to tell the difference by examining the time stamp, as discussed in the next section.

There are two parts to any METAR report: the body and the remarks. The body of a METAR report contains elements such as date/time, visibility, wind speed and direction, temperature, and altimeter setting; if present, elements in the body of a METAR

Synoptic Analysis and Forecasting. https://doi.org/10.1016/B978-0-12-809247-7.00003-X

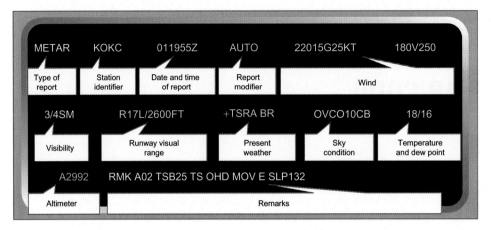

Fig. 3.1 Example METAR report from Oklahoma City, Oklahoma (KOKC). Everything before "RMK" is the body of the report, and the remarks section follows. From the Federal Aviation Administration: Aviation weather services (Advisory Circular, AC 00-45H), available online at: https://www.faa.gov/documentLibrary/media/Advisory_Circular/AC_00-45H.pdf.

are required to be reported. In addition, elements within the body of the METAR must always be reported in the same specific order. The remarks (RMK) section contains additional useful information that is not required and varies based on region and type of observation equipment.

3.1.1 Body of METAR

The body of a METAR report contains the following elements in order: station identifier, date and time, report modifier (if needed), wind, prevailing visibility, runway visual range (RVR, if needed), present weather, sky condition, temperature and dew point, and altimeter setting (Fig. 3.1). The RMK section, if present, immediately follows altimeter setting. Here, we use the METAR report from Oklahoma City, Oklahoma (KOKC) in Fig. 3.1 to explore the elements of the body of a METAR.

As mentioned previously, METAR or SPECI may be specified at the start of the report. Next is the four-letter station identifier code. These codes are based on the International Civil Aviation Organization (ICAO), which certifies airport identifiers. The first letter in the four-letter ICAO identifier is for the continent or subcontinent region, shown on the map in Fig. 3.2. For example, all continental US station identifiers begin with K, while all Canadian stations start with C. South American identifiers all start with S, while in Australia they have a leading Y. On continents with small countries such as Europe, the second letter of the identifier is

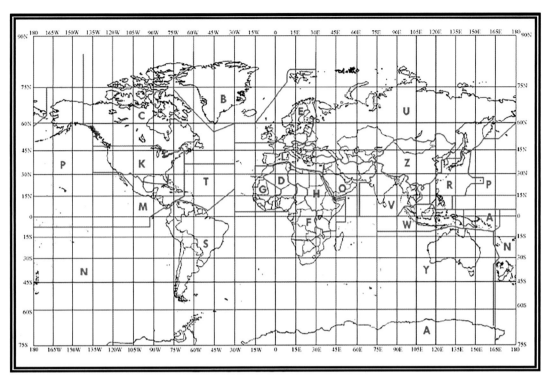

Fig. 3.2 Map of International Civil Aviation Organization (ICAO) continent and subcontinent codes, which compose the leading letter in the METAR ICAO station identifier group. From the Federal Aviation Administration: Aviation weather services (Advisory Circular, AC 00-45H), available online at: https://www.faa.gov/documentLibrary/media/Advisory_Circular/AC_00-45H.pdf.

typically unique to that country. In the contiguous United States, however, the last three letters of the station code are for the location, e.g., OKC for Oklahoma City.

The next element in the METAR is the date and time. The date is simply the two-digit day of the month, based on Zulu (Z) time. The four-digit time in Z immediately follows; METAR reports are always reported in Z, never in local time. This is primarily because METAR code was developed for aviation use, for which it is necessary that everyone operates on the same clock. Routine METAR reports will always be issued at the same time each hour. In most countries, this time is on the hour, e.g., 2000Z. However, in the United States, most stations report a few minutes before the top of the hour, as evidenced by the 1955Z in Fig. 3.1. Not all US stations are the same, either; routine reports can be issued anywhere between 51 and 59 min past the hour. It is important to: (a) recognize the regular hourly reporting time of a particular station(s) and (b) remember that routine hourly reports issued before the top of the hour count as the report for the next hour. For example, the 1955Z report in Fig. 3.1 is technically the 2000Z report.

After date and time is the report modifier, which is only included if necessary. The two options for the report modifier are AUTO and COR. AUTO specifies that the surface observation was taken by an ASOS with no human augmentation (Chapter 2). COR indicates that one or more elements of the report were corrected by a manual observer, typically shortly after report issuance. For example, perhaps the observer initially typed an incorrect digit in the wind speed and needed to issue a correction; COR indicates that the correction has been made and the METAR report is now accurate.

The next element of the METAR report is the wind. Recall that wind direction is always reported as the direction the wind is *coming from*. Wind direction is coded as three digits; in Fig. 3.1, the wind is *coming from* 220°, approximately southwesterly. Sustained wind speed is reported by the two (or in rare cases, three) digits that immediately follow wind direction, and the speed is always reported to the nearest whole kt. In Fig. 3.1, the sustained wind speed is 15 kt. Calm winds are reported as 00000KT, while sustained wind speeds >100 kt are reported as, e.g., 220110KT, as an extra digit is permitted for three-digit wind speeds. Now notice the G25 coded immediately after the sustained wind speed (Fig. 3.1); this is how wind gusts are reported to the nearest whole kt. A wind gust is only reported when the wind varies >10 kt between peaks and lulls during the past 10 min. In summary, the wind in Fig. 3.1 should be read as "sustained wind speed of 15 kt from 220° gusting to 25 kt."

Now notice the 180V250 immediately following the wind report (Fig. 3.1). This indicates that the wind direction has varied between 180° and 250°; it is only coded if the variation is >60° and the wind speed is >6 kt. If the wind is variable, but ≤6 kt, the METAR will have VRB coded instead of the three-digit wind direction. For example, a variable 5 kt wind is coded as VRB05KT.

Next in Fig. 3.1 is the *prevailing visibility* section, which in North America is typically reported in statute miles (SM), but often in meters on other continents. The prevailing visibility at KOKC is ¾SM; remember to not code a space between the number and SM. Also, it is important to understand that either the ASOS or a human observer determines prevailing visibility by assessing the farthest that can be seen across at least half the sky, not necessarily across a continuous sector. Finally, fully automated stations are not allowed to report visibility >10SM. Stations with manual observers will occasionally report ≥ 20 SM but must do so in increments of 5SM; this is most common at observation stations located at the top of large hills or mountains, such as Mount Washington in New Hampshire.

The next group is the RVR, which is only reported in certain locations and conditions. RVR reports the distance that is visible down a specific runway from a certain approach direction and is used almost exclusively by the aviation sector. The permanent R at the beginning of the group indicates the group heading. The number after the R is the runway number (17 in Fig. 3.1), followed immediately by the approach direction; R is for right, C for center, and L for left. After the slash is the visibility down the runway, which in the case of the KOKC report in Fig. 3.1 is 2600 ft. The minimum RVR that will be specified is 600 ft., while the maximum is 6000 ft. If RVR is <600 ft., it is reported as M0600, and when RVR is >6000 ft., it is written as P6000. The letters M and P are used as abbreviations for the French words meaning less than and greater than, respectively. Another possibility is that RVR is varying, in which case two values are written with a V in between. For example, RVR varying between 1200 and 6000 ft. on runway 14 from the left approach would be reported as R14L/1200V6000FT.

We now move to the present weather group in the METAR report, which in the KOKC report consists of "+TSRA BR." Present weather is where precipitation intensity, type, and obscurations to visibility, among other things, are reported. When decoding present weather, it is important to remember the order in which elements must be written within the group. Fig. 3.3 illustrates the proper order from left to right. The first element is precipitation intensity or, in some cases, distance. Light precipitation is denoted by a negative sign, moderate precipitation with no symbol, and heavy precipitation by a positive sign. When precipitation or obscuration of any intensity is in the vicinity, defined as 5–10SM from the observation station, a leading VC is added (Fig. 3.3). Recall that present weather is said to be occurring at a station if it is within 5SM of that location.

The second element of the present weather group is called the *descriptor*. The descriptor depends on the type of precipitation or obscuration, but TS (thunderstorm) and SH (shower) are generally the most common. Some descriptors such as DR (drifting) and BL (blowing) can only be used with specific precipitation or obscuration types such as snow and dust. The third element is precipitation, with all possibilities shown in Fig. 3.3. It is important to understand that although descriptors are written before precipitation, the intensity refers to the precipitation, not the descriptor. For example, +TSRA from the KOKC METAR

Qualifier				Weather phenomena				
Intensity or proximity		Descriptor		Precipitation		Obscuration		Other
1		2		3		4		5
−	Light	MI	Shallow	DZ	Drizzle	BR	Mist	PO	Dust/sand whirls
	Moderate[2]	PR	Partial	RA	Rain	FG	Fog	SQ	Squalls
+	Heavy	BC	Patches	SN	Snow	FU	Smoke	FC	Funnel cloud, tornado, or waterspout[4]
VC	In the vicinity[3]	DR	Low drifting	SG	Snow grains	VA	Volcanic ash	SS	Sandstorm
		BL	Blowing	IC	Ice crystals (diamond dust)	DU	Widespread dust	DS	Duststorm
		SH	Shower(s)	PL	Ice Pellets	SA	Sand		
		TS	Thunderstorms	GR	Hail	HZ	Haze		
		FZ	Freezing	GS	Small hail and/or snow pellets	PY	Spray		
				UP	Unknown precipitation				

1. The weather groups are constructed by considering columns 1 to 5 in the table above in sequence, i.e., intensity followed by description, followed by weather phenomena, e.g., heavy rain shower(s) is coded as +SHRA.
2. To denote moderate intensity no entry or symbol is used.
3. See text for vicinity definitions.
4. Tornadoes and waterspouts are coded as +FC.

Fig. 3.3 List of items that can be reported in the METAR present weather group. In order from left to right: precipitation intensity, descriptor, and type, obscurations to visibility, and other weather phenomena. From the Federal Aviation Administration: Aviation weather services (Advisory Circular, AC 00-45H), available online at: https://www.faa.gov/documentLibrary/media/Advisory_Circular/AC_00-45H.pdf.

(Fig. 3.1) should be read as "heavy rain with a thunderstorm," not "a heavy thunderstorm and rain."

The fourth element of the present weather group consists of obscurations such as FG (fog) and BR (mist). Some obscurations can have their own descriptors; for example, light snow with freezing fog would be coded as −SN FZFG. The fifth and final element ("other") is composed of relatively rare phenomena such as squalls, funnel clouds, tornadoes, and dust storms. Funnel clouds, which do not touch the ground, are coded as FC, while tornadoes and waterspouts, which do touch the ground, are coded as +FC. We will discuss several more examples of present weather METAR reports throughout the remainder of this chapter.

We now move to the sky condition group in the METAR, where sky coverage and cloud heights are reported along with any special cloud types occurring. As seen in Fig. 3.1, the format is a three-letter sky coverage amount, e.g., OVC for overcast, immediately followed by the three-digit cloud-base height given in hundreds of feet, e.g., 010 for 1000 ft. Any special cloud types such as cumulonimbus (CB), towering cumulus (TCU), or mammatus (MAM) will immediately follow the cloud base height.

The possibilities for sky coverage are listed in Fig. 3.4A. Recall from Chapter 2 that the ASOS or manual observer reports calculates sky coverage in eighths; this value is then converted into a METAR abbreviation such as OVC or BKN for broken. Multiple cloud base heights can be reported in a METAR; for example, BKN007 OVC010 should be read as broken skies at 700 ft. and overcast at 1000 ft. However, multiple cloud layers can only be reported with the summation rule in mind. That is, because all surface observations are taken from the ground, it is impossible to tell if the sky is less covered above a layer of a certain cloud coverage. For example, one cannot code OVC010 SCT015 because an observer at the ground cannot see above the overcast layer. Only the same or greater sky coverage as is present at lower levels can be reported. As examples, FEW007 BKN008 OVC010 and SCT050 SCT075 are permissible. Finally, recall that an ASOS can only detect clouds as high as 12,000 ft. Therefore, at fully automated (AUTO) stations, CLR will not account for the possibility of high (>12,000 ft.) cloud bases.

A ceiling, which is particularly crucial for the aviation industry, is defined as the lowest BKN or OVC layer. For example, BKN007 OVC010 indicates that the ceiling is at 700 ft.; in Fig. 3.1, the ceiling is 1000 ft. Now examine the VV contraction in Fig. 3.4. This stands for vertical visibility, which is a condition that is reported when the sky is obscured and the ceiling is undefined. This situation most commonly occurs during fog, when the ceiling cannot be discerned. The maximum height that can be seen upward is reported with VV. For example, VV006 means sky obscured with a vertical visibility of 600 ft. The difference between low overcast and vertical visibility is depicted in Fig. 3.4B. Vertical visibility is particularly hazardous to aviation because it reduces slant range visibility during takeoff and landing (Fig. 3.4B).

Reportable Contraction	Meaning	Summation Amount of Layer
VV	Vertical Visibility	8/8
SKC or CLR[1]	Clear	0
FEW[2]	Few	1/8 – 2/8
SCT	Scattered	3/8 – 4/8
BKN	Broken	5/8 – 7/8
OVC	Overcast	8/8

1. The abbreviation CLR will be used at automated stations when no layers at or below 12,000 feet are reported; the abbreviation **SKC** will be used at manual stations when no layers are reported.
2. Any layer amount less than 1/8 is reported as FEW.

(A)

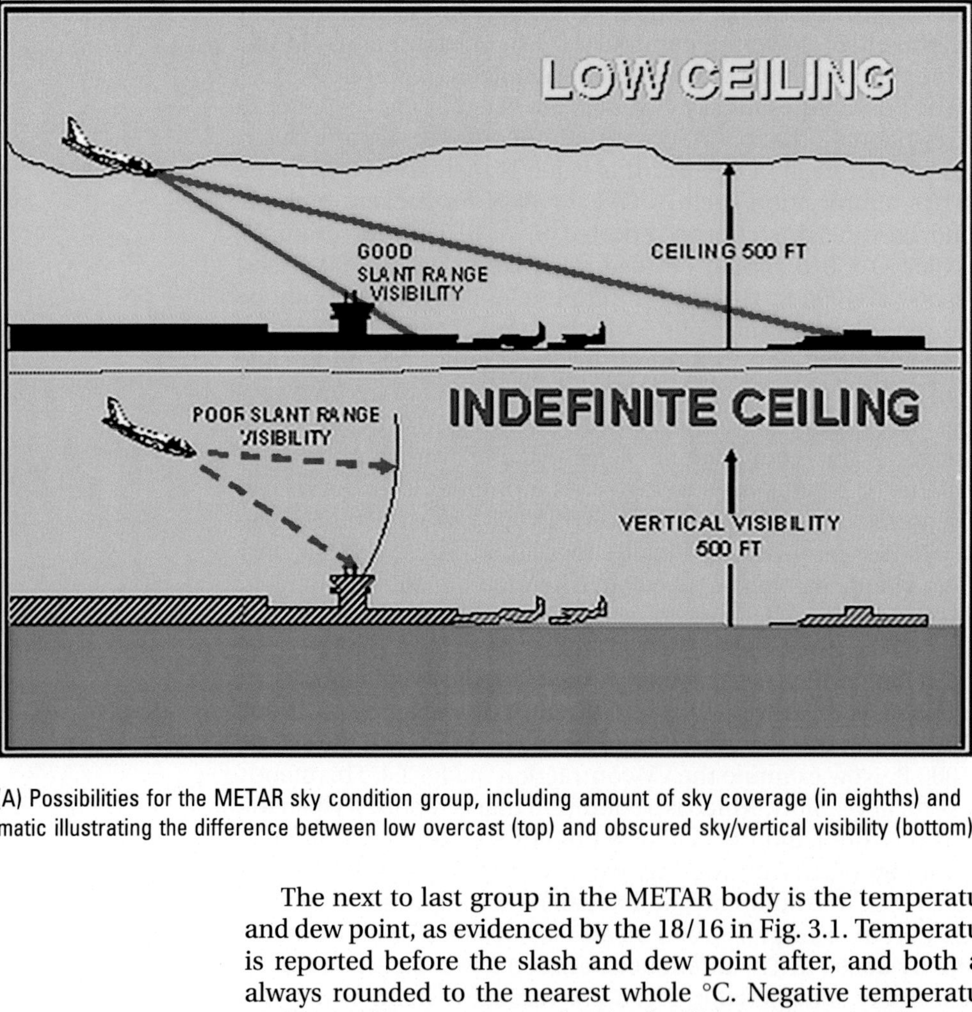

(B)

Fig. 3.4 (A) Possibilities for the METAR sky condition group, including amount of sky coverage (in eighths) and (B) schematic illustrating the difference between low overcast (top) and obscured sky/vertical visibility (bottom).

The next to last group in the METAR body is the temperature and dew point, as evidenced by the 18/16 in Fig. 3.1. Temperature is reported before the slash and dew point after, and both are always rounded to the nearest whole °C. Negative temperature and dew points are reported with a leading M. For example, a temperature of −3°C and a dew point of −5°C would be reported as

M03/M05. Both temperature and dew point are always reported in °C; to convert to US surface station plots (Chapter 2), the equivalent °F values must be calculated. In Fig. 3.1 example, we see that the temperature and dew point are quite close together, indicating a relatively large surface relative humidity.

The final group is *altimeter setting*, which in North America is always denoted by a leading A. Altimeter setting is sea-level pressure (SLP) calculated using the standard atmosphere chart and is crucial to aviation. The leading A is followed by a four-digit pressure value given in inches of mercury (inHg), a unit of pressure (Chapter 1) that is still used by the aviation sector. In Fig. 3.1, the altimeter setting is 29.92 inHg. Later in this chapter, we will discuss international differences regarding altimeter setting in METAR code; some countries report altimeter settings in hPa.

3.1.2 METAR Remarks

To introduce elements in the RMK section, we will use the four METAR reports shown in Fig. 3.5. It is important that the novice analyst or forecaster remember that the contents of the RMK section vary greatly by location and weather situation. Weather observers have a lot of latitude as to what can be included in the RMK section. In this text, we will review some of the more common RMK elements used in North America.

The first element that follows RMK in Fig. 3.5A,B,D is A02, which specifies that the observation system is an ASOS that can discern precipitation type. Older ASOS systems that cannot discern precipitation type are coded as A01. Neither of these codes necessarily means that the station is fully automated; a fully

(A) Birmingham, Alabama
   KBHM 171153Z 16012G23KT 8SM -RA BKN018 OVC026 16/13 A2965
   RMK A02 PK WND 16027/1140 RAB49 SLP035 70006 P0000 T01560128 10161 20150 58022 $
(B) Islip, New York
   KISP 131156Z AUTO 12014G26KT 9SM OVC007 22/21 A2969
   RMK A02 PK WND 14026/1154 RAE55 SLP053 P0148 61292 71310 T02170206 10217 20189 58013
(C) St Louis, Missouri
   KSTL 230124Z 33007G35KT 280V020 1 SM R30R/1600VP6000FT +FC -TSRA BR BK020 OVC040CB 19/17 A2976
   RMK TORNADO B10 W MOV E A02 PKWND 29042/0112 WSHFT 0104 CONS LTGICCGCC ALQDS TS ALQDS MOV E P0040
(D) Hutchinson, Kansas
   KHUT 100248Z 34023G32KT 2 1/2SM TS FZRAGS BR BKN010 BKN014 OVC 019 00/M02 A2992
   RMK A02 PK WND 33032/0246 LTGDSNT ALQDS TSE09B26UPB09E10RAB31E45FZRAE03B45SNB03E09GSB48 TS OHD MOV E P0014

Fig. 3.5 Example METAR reports from (A) Birmingham, Alabama (KBHM), (B) Islip, New York (KISP), (C) St Louis, Missouri (KSTL), and (D) Hutchinson, Kansas (KHUT).

automated station will be identified with AUTO in the body of the METAR (Section 3.1.1).

The next element is the peak wind, abbreviated PK WND. In the United States, this is defined as the maximum three-second wind speed during the past hour and is only reported if the wind speed is $\geq$25 kt. As in the body of the METAR, a three-digit wind direction is followed by a two- (or three-) digit wind speed. After a slash, the four-digit time at which the peak wind occurred is reported. In Fig. 3.5A, the peak wind was 27 kt from 160° and occurred at 1140Z. Another example is observed in Fig. 3.5D, where the peak wind was 32 kt from 330° at 0246Z.

We now move to the precipitation and thunderstorm start/end times element, which in Fig. 3.5A is written as RAB49, meaning rain began (B) at 49 min past the hour (1149Z). In Fig. 3.5B, we see that the rain ended at KISP at 1155Z (RAE55). If a start time is specified, but an end time is not, it means the precipitation is still occurring; this is the case at KBHM (Fig. 3.5A), where light rain is reported in the body of the METAR. A unique situation is seen at KHUT (Fig. 3.5D), where several types of precipitation and thunderstorms began and ended within the same hour, including "unknown precipitation" (UP), which an ASOS will sometimes report when mixed precipitation is occurring. In the body of the KHUT report, we see that a thunderstorm (TS), freezing rain (FZRA), small hail (GS), and mist (BR) are all reported, indicating very wild weather in central Kansas.

The next element in the RMK section is SLP. In Fig. 3.5A, we see SLP035, which translates to 1003.5 hPa. The format is the same as that of the SLP in surface station plots (Chapter 2), where the leading 9 or 10 is left off. For example, in Fig. 3.5B, we can deduce that the SLP is 1005.3 hPa, because 905.3 hPa is an unrealistic value except within very strong tropical cyclones. The SLP reported in the RMK section should *not* be considered a substitute for altimeter setting in the body of the METAR. Both are representative of SLP, but they are calculated differently. Altimeter setting is calculated using the standard atmosphere chart, while SLP in the RMK section is calculated using temperature information. Aviators should always use altimeter setting, while meteorologists typically find SLP in the RMK section to be more accurate for a given location.

The next set of elements in the RMK section detail precipitation accumulations. The P group in each report in Fig. 3.5 details the amount of precipitation that has fallen since the previous routine hourly METAR. Precipitation is reported in hundredths of an inch (in.); for example, at KISP (Fig. 3.5B), 1.48 in. has fallen in the past hour. At KBHM, we see P0000, which means that a trace of precipitation fell during the past hour. Trace amounts are <0.01 in., but should not be confused with zero precipitation; if no precipitation occurred, the P group will not be in the METAR.

Now examine the next group that begins with 6, seen at KISP (Fig. 3.5B); this is the 3- or 6-h precipitation accumulation, also reported in hundredths of an inch. Whether it is 3- or 6-h precipitation depends on the time of the METAR report. At 03Z, 09Z, 15Z, and 21Z, the 6 group reports 3-h precipitation totals; at 00Z, 06Z, 12Z, and 18Z, 6-h totals are coded. This group always starts with a 6 regardless of whether 3- or 6-h precipitation is being reported, so it is crucial to know the time of the METAR report. In the KISP flood situation shown in Fig. 3.5B, a record 12.92 in. of precipitation fell in the past 6 h; we know it is a 6-h total because it is the 12Z METAR report.

Temperature groups follow the P and 6 group and are quite useful for forecast verification or climatology records. In Fig. 3.5A and B, we see the T group, which is temperature and dew point reported to the nearest tenth of a °C, i.e., with increased precision relative to the body of the METAR. The number immediately following the T will always be 0 or 1; 0 indicates a positive temperature, while 1 indicates a negative temperature. The three subsequent digits are temperature. In Fig. 3.5A, the temperature is +15.6°C, while in Fig. 3.5B, it is +21.7°C. Following the temperature, dew point is reported using the exact same format. In the KBHM report, the dew point is +12.8°C, while in the KISP report it is +20.6°C.

Following the T group are the 1 and 2 groups (Fig. 3.5A and B), which represent the 6-h maximum and minimum temperatures, respectively. These groups only appear in 00Z, 06Z, 12Z, and 18Z routine METAR reports. The 1 group, in which the second digit is always a 0 or 1 for positive and negative, respectively, indicates the 6-h maximum temperature. The 2 group follows the exact same format, but for the 6-h minimum temperature. In the KBHM report, the maximum temperature over the past 6 h was +16.1°C, while the minimum was +15.0°C (Fig. 3.5A). In the KISP report, the maximum was +21.7°C, while the minimum was +18.9°C. These METAR groups are very useful for immediate forecast verification and climatology records.

Following the maximum and minimum temperature groups is the 5 group (Fig. 3.5A and B). This is the 3-h SLP tendency as discussed with respect to surface station plots in Chapter 2. Because it is always a 3-h SLP tendency, the 5 group only appears at 00Z, 03Z, 06Z, and so forth. The first digit immediately following the 5 is a special code taken from the table in Fig. 3.6B, which offers various descriptions of SLP trends. Each description should match

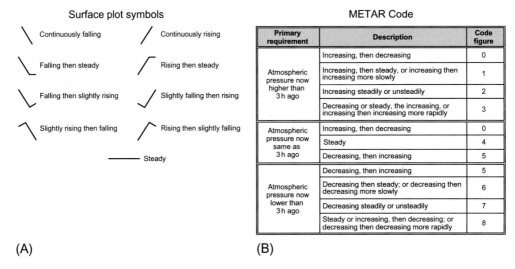

Surface plot symbols

Continuously falling	Continuously rising
Falling then steady	Rising then steady
Falling then slightly rising	Slightly falling then rising
Slightly rising then falling	Rising then slightly falling
Steady	

METAR Code

Primary requirement	Description	Code figure
Atmospheric pressure now higher than 3 h ago	Increasing, then decreasing	0
	Increasing, then steady, or increasing then increasing more slowly	1
	Increasing steadily or unsteadily	2
	Decreasing or steady, the increasing, or increasing then increasing more rapidly	3
Atmospheric pressure now same as 3 h ago	Increasing, then decreasing	0
	Steady	4
	Decreasing, then increasing	5
Atmospheric pressure now lower than 3 h ago	Decreasing, then increasing	5
	Decreasing then steady; or decreasing then decreasing more slowly	6
	Decreasing steadily or unsteadily	7
	Steady or increasing, then decreasing; or decreasing then decreasing more rapidly	8

(A) (B)

Fig. 3.6 For the SLP tendency group in the METAR remarks (RMK) section: (A) surface plot symbols from Chapter 2, and (B) the corresponding METAR code number that follows the 5 at the start of the SLP tendency group. From the Federal Aviation Administration: Aviation weather services (Advisory Circular, AC 00-45H), available online at: https://www.faa.gov/documentLibrary/media/Advisory_Circular/AC_00-45H.pdf.

with a surface station plot symbol in Fig. 3.6A and discussed in Chapter 2. The final three digits are the SLP tendency in tenths of hPa. For example, in the KBHM report, the SLP has decreased 2.2 hPa over the past 3 h, while at KISP, it has decreased 1.3 hPa. We know it is a decrease because of the second digit (8) in the 5 group (Fig. 3.6B).

The METAR RMK section can also feature any plain language that the observer deems necessary, such as the tornado description at KSTL (Fig. 3.6C), or thunderstorm and lightning information. For example, in the KSTL report, the observer is reporting constant (CONS) in-cloud (IC), cloud-to-ground (CG), and cloud-to-cloud (CC) lightning in all quadrants (ALQDS) around the station. A thunderstorm is also being reported in all quadrants. In general, the observer will include as much hazard frequency, distance, and type information as they can.

Finally, notice the dollar sign at the end of the KBHM report (Fig. 3.5A). This indicates that one or more sensors in the ASOS requires maintenance. When this sign is present, the METAR report should be questioned, especially if it is fully automated. It is not a guarantee, however, that something is incorrect in the METAR. In the case of the KBHM report, it is not a fully automated station, meaning we can generally trust the report even if minor or routine sensor maintenance is required.

3.1.3 Criteria for SPECI

SPECI can be issued up to 12 times per hour, in addition to the routine hourly METAR. However, there are specific criteria for SPECI to be issued, detailed in Fig. 3.7. It is important to remember that an ASOS or manual observer will issue a SPECI when one of the listed criteria occurs. The most common causes of SPECI reports are visibility or ceiling changing up or down through

1	Wind shift	Wind direction changes by 45° or more in less than 15 min and the wind speed is 10 knots or more throughout the wind shift
2	Visibility	Surface visibility as reported in the body of the report decreases to less than, or if below, increases to equal or exceed: a. 3 miles b. 2 miles c. 1 mile d. The lowest standard instrument approach procedure minimum as published in the National Ocean Service (NOS) US *Instrument Procedures*. If none published use ½ mile
3	Runway visual range (RVR)	The highest value from the designated RVR runway decreases to less than, or if below, increases to equal or exceed 2,400 feet during the preceding 10 min US military stations may not report a SPECI based on RVR
4	Tornado, funnel cloud, or waterspout	a. is observed b. disappears from sight, or ends
5	Thunderstorm	a. begins (a SPECI is not required to report the beginning of a new thunderstorm if one is currently reported) b. ends
6	Precipitation	a. hail begins or ends b. freezing precipitation begins, ends, or changes intensity c. ice pellets begin, end, or change intensity
7	squalls	When they occur
8	Ceiling	The ceiling (rounded off to reportable values) forms or dissipates below, decreases to less than, or if below, increases to equal or exceed: a. 3000 ft b. 1500 ft c. 1000 ft d. 500 ft e. The lowest standard instrument approach procedure minimum as published in the National Ocean Service (NOS) US *Instrument Procedures*. If none published, use 200 feet
9	Sky condition	A layer of clouds or obscurations aloft is present below 1000 ft and no layer aloft was reported below 1000 ft in the preceding METAR or SPECI
10	Volcanic eruption	When an eruption is first noted
11	Aircraft mishap	Upon notification of an aircraft mishap, unless there has been an intervening observation
12	Miscellaneous	Any other meteorological situation designated by the responsible agency of which, in the opinion of the observer, is critical

Fig. 3.7 All possible criteria for the issuance of a SPECI report. From the Federal Aviation Administration: Aviation weather services (Advisory Circular, AC 00-45H), available online at: https://www.faa.gov/documentLibrary/media/Advisory_Circular/AC_00-45H.pdf.

one of the thresholds listed in Fig. 3.7, dramatic wind shifts, and thunderstorms beginning or ending. Note that SPECI are only issued for the start or end of precipitation if the precipitation type is hail, freezing precipitation (i.e., freezing rain, freezing drizzle), or ice pellets; they are *not* issued for the start or end of rain or snow.

The two METAR reports in Fig. 3.5C and D are SPECI, not routine hourly reports. We know this because they were issued at 24 and 48 min past the hour, respectively, which are not the typical routine hourly reporting times. Although we cannot be certain about the exact reason for the SPECI issuances without examining the previous hour or two of METAR reports, we can infer that the reason for the SPECI in Fig. 3.5C is the tornado that was spotted west of KSTL moving toward the east. Soon after this SPECI during the 2011 St Louis Airport tornado, the weather observers had to move to shelter! The SPECI from KHUT (Fig. 3.5D) could have been issued for several reasons including ceiling and visibility changes or a wind shift. However, GS (small hail) begins at 48 min past the hour, the same time as the SPECI, making this the most likely reason for issuance.

3.1.4 International Differences

To this point we have only examined METAR reports from US stations. Although the elements in the body of a METAR are standard across the globe, the units in which some of these elements are reported can vary. The following is a list of key differences in the body of METAR reports from other regions as compared to those from North America:

- Wind speed reported in meters per second (MPS).
- Prevailing visibility and RVR reported in meters; a conversion table is included in Table 3.1.
- Inclusion of a prevailing visibility trend (up [U], down [D] or no change [N]).
- Altimeter setting in hPa, with a leading Q instead of A.
- CAVOK, meaning "clouds and visibility are okay."

How many of the international differences listed above are used depends on the country issuing the METAR report. Canada, for example, uses the same conventions as the United States in the body of the METAR, although the information included in the RMK section is frequently different. Fig. 3.8A shows a report from Sudbury, Ontario (CYSB). Notice that all elements in the body of the report are like a US METAR report, apart from the report being issued exactly on the hour. In the RMK section, the main difference from US reports is the inclusion of cloud type. This is

Table 3.1 For the Visibility Group in the Body of a METAR Report, Conversions Between Common Values of Visibility in Meters (International) and Statue Miles (North America)

Meters (m)	Statute Miles (SM)	Meters (m)	Statute Miles (SM)
0200	1/8	3200	2
0400	¼	3600	2 ¼
0800	½	4000	2 ½
1200	¾	4800	3
1600	1	6000	4
2000	1 ¼	8000	5
2400	1 ½	9000	6
2800	1 ¾	9999	6+

(A) Sudbury, Canada
CYSB 171800Z 13010KT 2SM -SN BR BKN005 OVC013 M02/M03 A2946
RMK SF6ST2 CLOUD TO THE NORTH LOOKS LIKE A PUPPY SLP996
(B) Moron, Spain
LEMO 291900Z 26008KT 220V280 CAVOK 35/15 Q1018
(C) Odiham, United Kingdom
EGVO 291944Z 22013KT 3000 RA BKN015 BKN090 14/13 Q1010

Fig. 3.8 Example METAR reports from (A) Sudbury, Canada (CYSB), (B) Moron, Spain (LEMO), and (C) Odiham, United Kingdom (EGVO).

denoted by SF6ST2, which translates to stratus fractus (scud clouds) covering 6/8 of the sky and stratus covering 2/8 of the sky. Added together, these constitute an overcast sky, which correspond to the body of the METAR (OVC013). After cloud type, we see that the human observer was having a bit of fun in discerning what the cloud to the north looked like; this is a great example of the virtually endless latitude an observer has in the remarks section!

Fig. 3.8B and C is reports from Spain and the United Kingdom, respectively. First, notice the different leading letters of the ICAO station code (Fig. 3.2). In the Spain report, CAVOK is used, indicating that clouds and visibility are okay; this is never written in a US or Canadian METAR. In the United Kingdom METAR, visibility is reported as 3000, indicating that the prevailing visibility is 3000 m. We can tell the difference between m and SM because visibility in m will always be a much larger number and SM is not coded immediately after the visibility value. Finally, altimeter setting in both Spain and the United Kingdom is reported in hPa with a leading Q. This is quite a common practice outside of North America and should not be confused with the SLP sometimes reported in the RMK section.

3.2 METAR Code: Examples

As in most aspects of weather analysis and forecasting, decoding and writing METARs becomes easier with repetition and practice. To that end, Fig. 3.9 shows four consecutive routine hourly METAR reports (only the body of the reports is included) from Salt Lake City, Utah (KSLC) during February of some year. In the 1853Z (1900Z) report, the prevailing visibility is 2SM with light freezing rain (−FZRA) and mist (BR). Clouds are overcast at 4400 ft. (OVC044), and the temperature is −2°C with a dew point of −4°C. The altimeter setting is a rather high 30.30 inHg.

At 1953Z, the largest change is the broken cloud layer at 700 ft. (BKN007), which means the ceiling has lowered substantially since 1853Z. At 2053Z, the wind is now calm (00000KT) and the precipitation has stopped. However, mist is still present and the ceiling is a very low 600 ft. (OVC006). Finally, at 2153Z, light precipitation has resumed and winds are variable at 5 kt (VRB05KT). The ceiling has lifted to 1100 ft. (OVC011), because a scattered cloud layer (SCT) does not constitute a ceiling.

KSLC 241853Z 36003KT 2SM -FZRA BR OVC044 M02/M04 A3030

KSLC 241953Z 35005KT 2SM -FZRA BR BKN007 OVC044 M02/M04 A3028

KSLC 242053Z 00000KT 2SM BR OVC006 M02/M03 A3026

KSLC 242153Z VRB05KT 2SM -RA BR SCT007 OVC011 M01/M03 A3026

Fig. 3.9 METAR reports for four consecutive hours at Salt Lake City, Utah (KSLC).

3.3 Questions and Practice Exercises

1. In Fig. 3.9, the 2153Z METAR contains an error, made by either the ASOS or manual observer. What is the error?
2. In the present weather section of a METAR, how would "heavy rain with a thunderstorm and fog" be coded? Be sure to write the code in the correct order.
3. In a 2100Z METAR report remarks section, P0182 and 60352 are reported. How much precipitation has fallen in the past hour? What about the past 6 h? Explain your answers.
4. Use the following SPECI from Daytona Beach, Florida (KDAB) to answer the subsequent questions.

    KDAB   251841Z   21019G34KT   1/2SM + TSRA   FEW007 BKN022CB OVC030 23/21 A2983 RMK AO2 PK WND 27034/1837 TSB32RAB35 TS ALQDS FRQ LTGC ALQDS

    a. Describe where thunderstorms and lighting are being reported, and the type of lightning being observed.
    b. What is the ceiling height?
    c. At what times did the thunderstorm begin and end?
    d. What are the temperature and dew point?
5. For the SPECI in Question 4, draw a surface station plot based on the plotting conventions and procedures discussed in Chapter 2.
6. Use the following routine METAR from Boston, Massachusetts (KBOS) to answer the subsequent questions.

    KBOS 150354Z 04026G37KT 1 3/4SM RA BR OVC008 05/04 A2963 RMK AO2 PK WND 04039/0336 SFC VIS 2 RAB43 SLP034 P0010 T00500044

    a. Translate the reported wind direction and speed, including gusts and peak wind.
    b. What is the prevailing visibility? Was it reported from the tower or the surface? How do you know?
    c. What is the present weather being reported?
    d. How much precipitation fell in the past hour?

4

UPPER-AIR OBSERVATIONS

4.1 Introduction to Radiosondes

4.1.1 Radiosonde Instrumentation

The radiosonde[1] (rawinsonde) was invented in the late 1920s by Vilho Vaisala in Finland and independently by Pavel Molchanov in the Soviet Union. Radiosondes revolutionized our ability to measure basic atmospheric variables above the surface throughout the troposphere and stratosphere; previously, upper-air observations were expensive and quite sporadic. In modern times, these observations also serve as a crucial source of input data for numerical weather prediction (NWP) models. While the contents of the instrumentation package can vary slightly, standard radiosondes measure vertical profiles of temperature, humidity, and pressure. Vertical profiles of wind speed and direction are typically inferred from weather balloon drift, using radiosonde trackers such as radio direction finders or Global Positioning Systems (GPS). Technically, the term rawinsonde[2] is reserved for weather balloons with instrumentation that can determine wind information, although in practice, radiosonde and rawinsonde are frequently used interchangeably.

The radiosonde instrumentation is contained within a small white box attached to a weather balloon that is filled with either hydrogen or helium (Fig. 4.1). Fig. 4.2 shows a close-up photo of typical radiosonde instrumentation. Similar to its ASOS counterpart (see Chapter 2), the temperature sensor uses the principle of electrical resistance to measure temperature changes with altitude. The humidity sensor measures changes in ambient water vapor content by using substances that respond to such changes. Similarly to the ASOS pressure sensors (see Chapter 2), an aneroid

[1] The official American Meteorological Society (AMS) Glossary definition of a radiosonde is available online at: http://glossary.ametsoc.org/wiki/Radiosonde.
[2] The official American Meteorological Society (AMS) Glossary definition of a rawinsonde is available online at: http://glossary.ametsoc.org/wiki/Rawinsonde.

Synoptic Analysis and Forecasting. https://doi.org/10.1016/B978-0-12-809247-7.00004-1

Fig. 4.1 Example of a radiosonde being launched with (bottom) small white box containing the standard instrumentation, (top) hydrogen or helium-filled weather balloon and (middle) orange parachute. From the National Oceanic and Atmospheric Administration (NOAA) Earth Systems Research Laboratory (ESRL), available online at: https://www.esrl.noaa.gov/gmd/obop/mlo/programs/esrl/ozonesondes/img/img_launching_sonde_hilo_1.JPG.

barometer is used to measure vertical changes in pressure. Finally, the transmitter (Fig. 4.2) uses radio signals to relay real-time radiosonde data back to a surface receiving system.

The entire radiosonde instrumentation package is powered by a small battery located inside the white box (Fig. 4.2). Although it varies, a typical balloon can rise up to 30 km above the surface, which is approximately the 10 hPa pressure level in the stratosphere. An average balloon rises at a rate of 300 m per min and, depending on winds, can drift as much as 200 km from the initial release point. Radiosonde instruments are generally considered accurate within 1°C of temperature, 2 hPa of pressure, and 5% relative humidity. Note that the instrumentation package is attached to a small parachute (Fig. 4.1); when the weather balloon finally bursts at high altitudes, the instrumentation package slowly floats back to the surface. However, only about 20% of radiosondes are recovered, meaning that a global network is a relatively expensive endeavor.

4.1.2 Global Radiosonde Network

Even with dramatic advances in remote sensing (e.g., satellite) technology, radiosonde data is still considered to be a crucial source of input to NWP models. Radiosonde data also provides

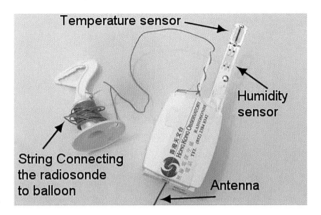

Fig. 4.2 Close-up photo of the standard radiosonde instrumentation, as indicated. The aneroid barometer is located inside the white box. From the Hong Kong Observatory, available online at: http://www.hko.gov.hk/prtver/html/docs/wservice/tsheet/uamet.shtml.

regular three-dimensional observations of the troposphere, which can be incredibly useful for forecasters.

The global radiosonde network is composed of more than 900 stations, including 92 in the United States and its territories. Each station is required to launch weather balloons twice a day, every day of the year, at 00Z and 12Z. This ensures that radiosonde data is gathered at the same times each day for every station across the globe. Intermediate (e.g., 06Z, 18Z) balloon launches are generally reserved for high-impact weather situations, such as a landfalling tropical cyclone or a severe weather outbreak.

Fig. 4.3 shows the location of radiosonde (sounding) stations across most of North America. There are a few important points to consider when examining this chart:

- Only a few stations are located over the oceans, which make up two-thirds of the global surface area.
- Spatial coverage on land is irregular and largely dependent on country. As examples:
  - The United States has a relatively high-density network, while Canada, which has a larger surface area, has a considerably more sparse radiosonde network (Fig. 4.3).
  - Northern Europe, China, and Japan have dense radiosonde networks (not shown), while less affluent countries in the same regions do not.
  - Less wealthy continents such as South America and Africa (not shown) have fewer radiosonde stations than other regions.

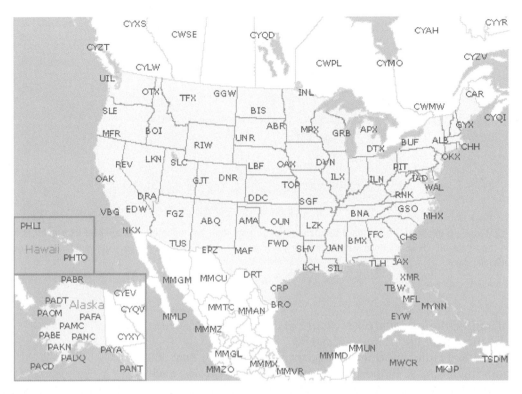

Fig. 4.3 Maps of current radiosonde stations in North America, excluding northern Canada and the eastern Caribbean. Each station is labeled using its International Civil Aviation Organization (ICAO) identifier code, although the leading "K" is left off for US stations. From the National Center for Atmospheric Research (NCAR) real-time weather data portal, available online at: http://weather.rap.ucar.edu/upper/.

- Remotely populated areas such as mountain ranges and polar latitudes have few sounding stations. Even in the United States, which has a dense network overall, there are fewer stations in the intermountain west than in other parts of the country (Fig. 4.3). These differences are even more exaggerated in places such as the Andes and Himalayas (not shown).
- Balloons can drift up to 200 km from their initial release point, depending on wind direction and speed. As such, soundings are typically considered valid for a 200-km radius around the launch location. Meteorologists sometimes refer to these as *proximity soundings.*

In all, the spatial and temporal limitations of the radiosonde network are its biggest drawback. In an ideal world, radiosonde launches would be more frequent and at more locations, but this would be quite cost-prohibitive, particularly in developing nations. In recent years, much of the gap in the radiosonde network has been filled by remote sensing data, particularly satellites. However, satellites still have accuracy issues in the lower troposphere (boundary layer), meaning that radiosondes are still an important aspect of the global observation network. As satellite technology improves, the global radiosonde network is likely to eventually become a thing of the past. However, the ability to plot and interpret atmospheric soundings (regardless of instrumentation or source) will remain a vital part of weather analysis and forecasting.

4.2   Upper-Air Station Plots

Meteorologists use radiosonde data to create and interpret upper-air charts that help with weather analysis and forecasting. Before learning how to plot and interpret these charts in Chapter 5, it is important to understand how upper-air station data is plotted. In Chapter 2, we discussed the surface station plot model; here, we discuss the upper-air station plot model, which is similar, but not identical.

Fig. 4.4 shows an example of a 500-hPa station plot using standard plotting conventions, with the data derived from a radiosonde sounding. Plotted at the top left is the 500-hPa temperature, in °C; unlike surface plots which vary by country, all upper-air temperatures and dew points are plotted in °C, regardless of pressure level or location. In the lower-left corner of Fig. 4.4 is the 500-hPa dew point depression, which is defined as temperature minus dew point. Because the dew point can never be warmer than the temperature, the dew point depression can never be a negative number. It is standard practice to plot dew point depression instead of dew point in upper-air station plots, in

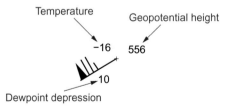

Fig. 4.4 Example of an upper-air station plot at 500 hPa. Counterclockwise from top left: temperature, dew point depression, and geopotential height. Wind information is plotted at the center of the plot, using the same conventions as surface station plots (see Chapter 2). Temperature and dew point depression are plotted in °C, wind speed in kt, and geopotential height in decameters (dam).

order to more readily identify regions of moisture and potential clouds; this will be further discussed in Chapter 5. In the Fig. 4.4 example, the temperature is −16°C and the dew point is −26°C.

Wind speed and direction are plotted at the center of the upper-air station plot, using conventions that are identical to surface station plots (see Chapter 2). Recall that, in the Northern Hemisphere, the wind speed flags and barbs are always plotted on the right-hand side of the flag stick looking outward, while in the Southern Hemisphere they are plotted on the left-hand side. As such, we know that the 500-hPa wind in Fig. 4.4 is in the Northern Hemisphere and would be reported as southwesterly at 75 kt. Finally, in the upper right of the upper-air station plot is the geopotential height, plotted in decameters (dam, 1 dam = 10 m). In Fig. 4.4, the geopotential height is 556 dam (5560 m). As we will see in Chapter 5, low (high) height on a constant pressure (isobaric)—e.g., 500 hPa—chart is analogous to low (high) pressure on a constant height (e.g., sea-level pressure) chart.

As we will see in Chapter 5, a map of upper-air station plots on a given pressure level (isobaric chart) can be used as a starting point for identifying upper-tropospheric weather systems and regions of clouds and precipitation, among other features. Therefore it is important for the analyst or forecaster to be fluent in decoding such plots. Fig. 4.5 shows examples of station plots from three commonly used pressure levels: 250, 500, and 700 hPa. All three plots use the standard upper-air station plot syntax, as shown in Fig. 4.4. For example, in the 250-hPa station plot, the temperature is −53°C, the dew point is −74°C, winds are from the north-northeast at 100 kt, and the geopotential height is 1044 dam (10,440 m). Now see if you can similarly decode the 500- and 700-hPa station plots.

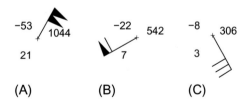

Fig. 4.5 Example of an upper-air station plots at (A) 250 hPa, (B) 500 hPa, and (C) 700 hPa. As in Fig. 4.4, plotted in each are wind direction and speed (kt), temperature and dew point depression (°C), and geopotential height (dam).

4.3 Questions and Practice Exercises

1. How many times a day are radiosondes typically launched and at what specific times (in Z) do launches occur each day?
2. On June 2nd of some year, a radiosonde is launched from Stony Plain, Alberta (near Edmonton) at noon Mountain Daylight Time. Is this a standard radiosonde launch? Explain how you know.
3. Name and explain two benefits and two limitations of the global radiosonde network.
4. Using standard Northern Hemisphere conventions, draw upper-air station plots for the following parameters at each pressure level.
   (a) 850 hPa:
   - Temperature: 12°C.
   - Dew point: 8°C.
   - Wind: Southeast at 17 kt.
   - Geopotential height: 1560 m.
   (b) 500 hPa:
   - Temperature: −7°C.
   - Dew point depression: 11°C.
   - Wind: North at 33 kt.
   - Geopotential height: 534 dam.
   (c) 300 hPa:
   - Temperature: −33°C.
   - Dew point: −52°C.
   - Wind: West at 97 kt.
   - Geopotential height: 1057 dam.
5. Draw an upper-air station plot for the parameters at in Question 4b) using standard *Southern Hemisphere* plotting conventions.

5

ISOPLETHS AND UPPER-TROPOSPHERIC CHARTS

5.1 Isopleth Rules and Conventions

Analyzing weather charts is one of the most important tasks that a meteorologist performs. While computer analysis has come a long way, manual chart analysis is still an invaluable exercise. In Chapters 2 and 4, we reviewed how to read surface and upper-air station plots, respectively. Over the next two chapters, we will discuss how to properly analyze upper- and lower-tropospheric charts, as well as how to identify key meteorological features on each chart. To do this, we will use *isopleths*, which are lines (contour) of a constant value of a meteorological variable (e.g., pressure, temperature). Table 5.1 shows specific isopleth names for common meteorological variables. For example, a line of constant pressure is called an *isobar*, while a line of constant temperature is an *isotherm*.

Drawing isopleths on a chart is a skill that improves rapidly with practice and repetition. While line colors and styles (i.e., solid vs. dashed) are largely individual choices, it is best practice to first draw isopleths using erasable pencil. Even the most experienced analyst can still make mistakes. To get started, Fig. 5.1 shows a chart of surface temperatures across the United States. For the sake of simplicity, temperatures in Fig. 5.1 are rounded to the nearest 10°F. The objective is to draw isotherms every 10°F. First, evaluate what the smallest and largest isotherms should be on the chart, which in our example are 20°F and 90°F, respectively. Next, choose an isotherm value, and start to connect points of equal value along that isotherm. It is best practice to begin at one end of the map and go toward the other, typically from left to right. In Fig. 5.1B, the 40°F isotherm starts in Washington State, then moves southeastward through Nevada, southern California,

Synoptic Analysis and Forecasting. https://doi.org/10.1016/B978-0-12-809247-7.00005-3

Table 5.1 Common Meteorological Variables and Their Corresponding Isopleth Names

Meteorological Variable	Isopleth Name
Pressure	Isobar
Temperature	Isotherm
Geopotential height	Isohypse (height contour)
Dew point	Isodrosotherm
Wind speed	Isotach
Pressure change	Isallobar
Thickness	Thickness line

southern Arizona, and into central Texas. To complete the 40°F isotherm, continue northeastward through Missouri, Illinois, and the Great Lakes, ending in Maine (Fig. 5.1C).

Fig. 5.1C shows the complete set of isotherms for our example. Note that all isotherms on the chart are clearly labeled. A few other important rules for drawing isopleths can be deduced from this exercise:

- Isopleths of a particular variable must never cross. Because only isotherms are plotted in Fig. 5.1C, no two lines can cross each other.
- Isopleths can only be drawn where there is representative data. In Fig. 5.1C, there are no isotherms that extend far into either ocean, Canada, or the Gulf Mexico; these are all regions where no temperature data is plotted.
- For a particular isopleth, values should be larger on one side of the isopleth and smaller on the other side. Examine the 40°F isotherm in Fig. 5.1C: temperature values are colder on one side, and warmer on the other side.
- Multiple isopleths of the same value *are* permitted on a single chart. In Fig. 5.1C, there are two 50°F isotherms, one on the west coast, and one extending from Texas to Massachusetts. Drawing multiple isotherms is much preferable to extending a single 50°F across regions lacking data, i.e., over Mexico and the Pacific Ocean in Fig. 5.1.
- Isopleths should be as smooth as possible, avoiding jagged edges or turns.

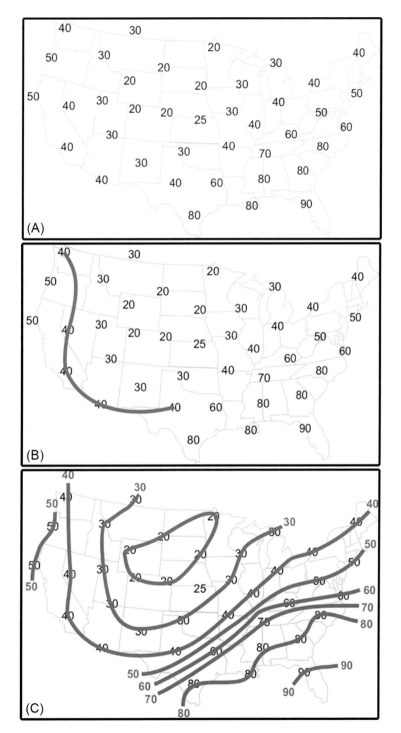

Fig. 5.1 Simplistic example of how to draw isopleths, using surface temperature data (°F): (A) US map of surface temperatures, (B) the 40°F isotherm plotted approximately halfway across the United States, and (C) a complete set of isotherms every 10°F between 20°F and 90°F, inclusive. Modified from the National Weather Service (NWS) JetStream: Learning lesson: Drawing conclusions—surface temperature map, available online at http://www.srh.noaa.gov/jetstream/synoptic/ll_analyze_temp.html.

5.2 Isopleths on Upper-Tropospheric Charts

For the remainder of this chapter, we will focus on isopleths on upper-tropospheric charts and discuss how to use them to identify important meteorological features. Table 5.2 shows isopleth conventions for 500-hPa and jet stream level (i.e., 200/250/300-hPa) charts. Recall that on an upper-air station plot, geopotential height is given in decameters (dam), where 1 dam = 10 m (e.g., 540 dam = 5400 m), and temperature is in °C, regardless of location. On 500-hPa charts, it is standard to plot geopotential height contours (isohypses) every 6 dam centered on 540 dam (i.e., 534, 540, 546, etc.), and isotherms every 5°C centered on 0°C (Table 5.2). On jet stream charts (Table 5.2), isohypses are plotted every 12 dam centered on 1020 dam (i.e., 1008, 1020, 1032 dam, etc.), temperature is typically not plotted, and isotachs (lines of constant wind speed) are plotted every 20 kt in order to outline jet streams and jet streaks.

5.2.1 500-hPa Charts

A 500-hPa isobaric chart is perhaps the most important upper-air plot for weather analysis and forecasting. Using these charts, we can identify upper-tropospheric troughs and ridges, which are very important for surface cyclone and anticyclone development, and weather forecasting. In this section, we will identify troughs and ridges by drawing isohypses across North America on the chart in Fig. 5.2. For the sake of simplicity, only geopotential height is plotted at each radiosonde station in Fig. 5.2; temperature, dew point depression, and wind, typically standard on upper-air station plots, are not shown. The geopotential height values in Fig. 5.2 are from 00Z on 13 March 1993, during the development of the intense midlatitude cyclone that would later be

Table 5.2 Standard Isopleth Conventions for Upper-Tropospheric Charts

Chart Type	Geopotential Height	Temperature	Other
500 hPa	Every 6 dam, centered on 540 dam	Every 5°C, centered on 0°C	None
200/250/ 300 hPa	Every 12 dam, centered on 1020 dam	None	Isotachs: every 20 kt, starting at either 40 or 60 kt

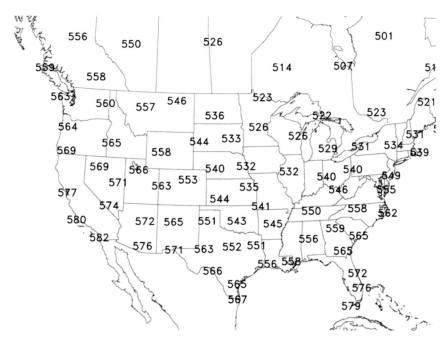

Fig. 5.2 Radiosonde-measured values of 500-hPa geopotential height (dam) across North America at 00Z 13 March 1993.

known as the 1993 "Superstorm.[1]" The 1993 Superstorm broke and still holds many of the nontropical cyclone low pressure records along the US east coast.

In order to isopleth Fig. 5.2, we first need to identify the smallest and largest isohypse values. Recall that 500-hPa geopotential height is contoured every 6 dam, centered on 540 dam (Table 5.2). In Fig. 5.2, the smallest geopotential height values are located in northern Ontario and Quebec. The smallest station value is 501 dam, meaning that the smallest isohypse on the chart will be 504 dam. The largest geopotential height values are located across southern California, specifically 582 dam at the station near San Diego. However, because only one station is reporting 582 dam and there are no values >582 dam, 576 dam will be our largest isohypse value. The requirement that at least two stations of a particular isopleth value or greater are necessary to draw an isopleth may differ among analysts and instructors; some prefer to draw an isopleth if there is just a single station reporting a particular value.

Fig. 5.3A shows the 504 dam isopleth across James Bay and northern Quebec. Because 504 dam is halfway between the 501

[1] Kocin, P. J., P. N. Schumacher, R. F. Morales, Jr., and L. W. Uccellini, 1995: Overview of the 12–14 March 1993 Superstorm. *Bull. Amer. Meteor. Soc,* **76,** 165–182.

and 507 dam station values, the 504 dam isohypse should be drawn precisely halfway between those two stations. In northern Quebec, however, 504 dam is a lot closer in value to 501 dam than to 510 dam, so the isohypse proximity should reflect that proportionality. Note that the 504 dam contour is drawn across a relatively small portion of the chart; isopleths must be reflective of the data and should only be extended as far as there are smaller values on one side of the isopleth and larger values on the other side.

Fig. 5.3B shows two separate 576 dam geopotential height contours, in addition to the aforementioned 504 dam line in Ontario and Quebec. From west to east, the first 576 dam contour starts just north of the Bay Area in northern California (577 dam station value) and ends in Tucson, Arizona (576 dam station value). We cannot extend the 576 dam isohypse into Mexico because although values to the north are smaller, the chart does not show data in Mexico (Fig. 5.2) and thus we cannot be certain that values to the south are larger. The second 576 dam contour starts north of Key West, Florida and goes through the 576 dam station value at West Palm Beach, Florida. We cannot extend this isohypse any further east or west because of the lack of data over water.

Fig. 5.3C shows the complete set of isohypses. Some height contours, such as 552 and 558 dam, go from coast to coast, while others such as 510 and 516 dam are limited to smaller regions. In this case, smaller geopotential height values are located in eastern North America and larger values are in western North America. As we will see in Section 5.3.1, this is indicative of a 500-hPa trough in the east and a ridge in the west; the trough in the east was associated with the formation and intensification of the 1993 Superstorm.

5.2.2 Jet Stream Charts

Now we examine 250-hPa charts, located near the top of the troposphere, or tropopause, where the jet stream is typically the strongest. Tropopause height varies with latitude and season, because the depth of the troposphere is proportional to its temperature. In summer, the troposphere is warmer and less dense, placing the tropopause at a higher altitude (lower pressure level). In the winter, the troposphere is colder and denser, meaning that the tropopause is located at a lower altitude (higher pressure level). In summer and in the tropics, the 200-hPa chart is typically most representative of the tropopause and jet stream height. In the winter and in polar latitudes, 300 hPa is typically more representative of the tropopause. As a compromise here, we will use

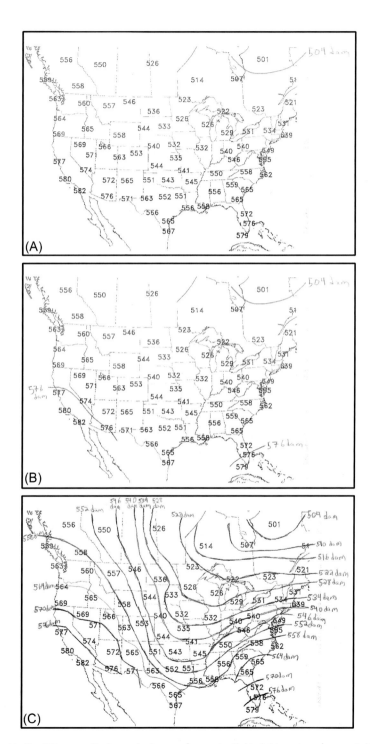

Fig. 5.3 For the chart in Fig. 5.2, manually drawn isohypses (geopotential height contours) every 6 dam: (A) the smallest geopotential height value (504 dam) on the chart, (B) the smallest and largest (504 and 576 dam) geopotential height value on the chart, and (C) the complete chart.

250-hPa charts, which usually can capture the jet stream throughout the year in the midlatitudes.

Upper-tropospheric troughs and ridges are still evident on a 250-hPa chart. However, the primary objective on these charts is to identify *jet streaks*, which are cores of fast winds within the jet stream. To isolate jet streaks, we will plot isotachs (lines of constant wind speed) on the blank chart in Fig. 5.4. On Fig. 5.4, both geopotential height and wind are plotted at 00Z 13 March 1993, the same time as our 500-hPa chart (Fig. 5.2). Recall from Chapter 4 that for station plots of wind speed, a flag represents 50 kt, a long barb is 10 kt, and a short barb is 5 kt. For example, at Bismarck, North Dakota, the 250-hPa geopotential height is 990 dam and the wind is 75 kt from the northwest (Fig. 5.4).

Fig. 5.5A shows 250-hPa isohypses every 12 dam (Table 5.2) across North America. The smallest value is 960 dam located in Ontario and Quebec, and the largest is 1068 dam: one 1068 dam contour extending from central California to southern Arizona, and the other across southern Florida. As on the 500-hPa chart, the smallest geopotential height values are located in the east and the largest are in the west.

Drawing isotachs is a bit different than drawing geopotential height contours. First, because we are trying to isolate the fastest winds on the chart, a reasonable minimum isotach value should be chosen. This value depends on latitude and season and is best judged on a case-by-case basis. For the cool season in the midlatitudes, 40 or 60 kt is generally a good choice. In our example in Fig. 5.5, we use 60 kt as our minimum value, with a contour interval of 20 kt. Second, one should plot the largest isotach value first, followed by progressively smaller isotachs. Third, isotachs should generally enclose wind speeds that are faster than the value of the isotach, whereas wind speeds slower than a particular isotach should be located outside of that isotach. Finally, regions inside an isotach are typically shaded, with a different colour used for each, as seen in Fig. 5.5B.

Fig. 5.5B shows a complete 250-hPa isotach chart. The 180-kt isotach encloses the station in western Pennsylvania at which a 185-kt wind speed is reported. As with isohypses and isotherms, the spacing of each isotach between stations should be proportional to the station values. In other words, the 180-kt isotach should be drawn closer to a station reporting 185 kt than a station reporting 165 kt. More than one isotach of the same value can be drawn on a single chart; such a situation is indicative of more than one jet streak.

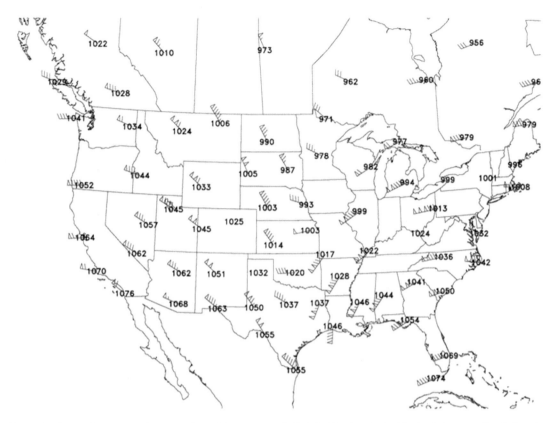

Fig. 5.4 Radiosonde-measured values of 250-hPa geopotential height (dam) and wind (kt, barbs) across North America at 00Z 13 March 1993.

From Fig. 5.5B, two jet streaks are evident, one centered in Ohio/Pennsylvania and the other oriented from northwest to southeast across the central Rockies. The eastern jet streak has faster maximum winds than the western jet streak, which we relate to the development of the 1993 Superstorm in Chapter 7. Although most isotachs are fully closed, the smallest ones (e.g., 60 kt) may not be able to be closed off. As with any type of isopleth, the analyst should draw what best represents the available data.

5.3   Feature Identification

5.3.1   Troughs and Ridges

Recall that friction is negligible above the atmospheric boundary layer (lowest ~1 km). Therefore for upper-tropospheric straight flow on the synoptic scale, the Pressure Gradient Force

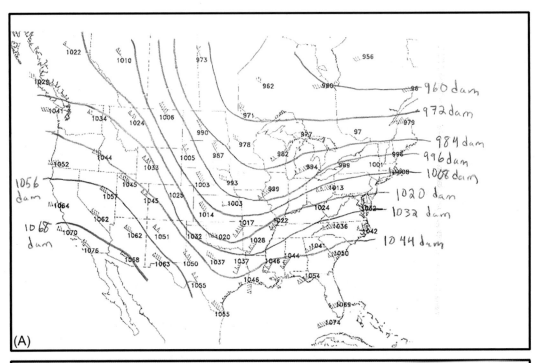

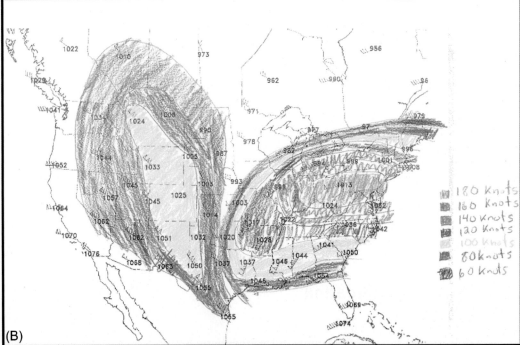

Fig. 5.5 For the chart in Fig. 5.4, manually drawn isohypses (geopotential height contours) every 12 dam and isotachs (every 20 kt starting at 60 kt): (A) the complete set of isohypses only and (B) the complete set of isotachs only.

(PGF) and Coriolis Force are assumed to be equal but opposite, a principle which is called *geostrophic balance*.[2] When the upper-tropospheric flow is curved, the Centripetal (Centrifugal) Force acts as a third force, and the resulting equilibrium is called *gradient balance*.[3] Regardless of whether the upper-tropospheric flow is straight or curved, we can diagnose the geostrophic or gradient wind direction by recalling that in the Northern Hemisphere, the wind is to your back with low height (pressure) on your left. In the Southern Hemisphere, with the wind at your back, low height (pressure) is on your right. This relationship between height/pressure and the geostrophic wind is called Buys Ballot's law.[4] Geostrophic and gradient winds are always parallel to geopotential height contours and have been shown to be at least 90% accurate with respect to the real wind above the boundary layer. Unless otherwise specified, for the remainder of this textbook we will assume that above the boundary layer, the wind blows parallel to isohypses.

On upper-tropospheric height charts, *troughs* are axes of geopotential height minima, while *ridges* are axes of geopotential height maxima. Geopotential height troughs and ridges are analogous to valleys and peaks in terrain, respectively. Fig. 5.6 shows schematics of a 500-hPa trough and ridge; trough axes are marked with a dashed line and ridge axes with a jagged line. Trough axes should be drawn through the line of maximum positive curvature (Fig. 5.6A), such that geopotential heights on either side of the trough axis are larger than along the axis itself. Similarly, ridge axes should be drawn through the line of maximum negative curvature (Fig. 5.6B), such that geopotential heights on either side are smaller than along the ridge axis. As we will discuss in Chapter 7, trough and ridge orientation are important for weather forecasting. A trough or ridge in the Northern Hemisphere is *positively tilted* when its axis has a positive slope, i.e., oriented from southwest to northeast, as both features are in Fig. 5.6. A *negatively tilted* trough or ridge features a negatively sloped axis, which in the case of a typical Northern Hemisphere feature would be oriented from southeast to northwest.

[2] For a full review of geostrophic balance and geostrophic wind, the reader is referred to University of Illinois, 2017: Weather World 2010, available online at: http://ww2010.atmos.uiuc.edu/(Gh)/guides/mtr/fw/geos.rxml.

[3] For a full review of gradient balance and gradient wind, the reader is referred to University of Illinois, 2017: Weather World 2010, available online at: http://ww2010.atmos.uiuc.edu/(Gh)/guides/mtr/fw/grad.rxml.

[4] American Meteorological Society, 2012: Glossary of Meteorology, available online at: http://glossary.ametsoc.org/wiki/Buys_ballot%27s_law.

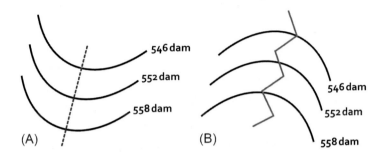

Fig. 5.6 Schematics of a Northern Hemisphere 500-hPa (A) trough (*dashed blue line*) and (B) ridge (*jagged red line*), using standard plotting symbols. Geopotential height contours are plotted in dam.

Fig. 5.7 shows a computer-analyzed geopotential height chart for the data in Fig. 5.2. Compare the computer analysis to the hand-drawn analysis in Fig. 5.3 and also to your own isopleth analysis; they should be similar! As expected, the primary trough in Fig. 5.7 is located in eastern North America, positively tilted from James Bay in northern Ontario southwestward through Oklahoma. Larger geopotential heights are located throughout western North America, and several ridge axes are noticeable, with one negatively tilted ridge just off the California coast marked in Fig. 5.7. Although we did not plot temperature on our 500-hPa

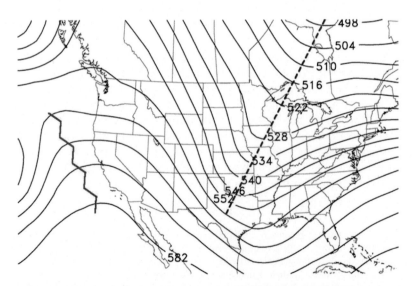

Fig. 5.7 Computer analysis of 500-hPa geopotential height (dam, *solid black contours*) across North America at 00Z 13 March 1993. An example trough (*dashed blue line*) and ridge (*jagged red line*) are marked using standard plotting symbols.

chart, troughs are regions of upper-tropospheric cold air, while ridges represent regions of upper-tropospheric warm air. We will expand on these concepts in Chapter 7.

5.3.2   Zonal and Meridional Flow

In addition to being able to recognize straight (geostrophic) vs. curved (gradient) flow, it is useful to differentiate between zonal and meridional flow. *Zonal flow* is defined as west to east or east to west, while *meridional flow* indicates north-to-south or south-to-north air motion. Zonal flow is typically associated with relatively weak troughs and ridges, and thus relatively quiet weather. Meridional flow is associated with more amplified (stronger) troughs and ridges, indicative of the poleward transport of heat and the equatorward transport of cold air. The more amplified the troughs and ridges are (i.e., the more meridional the flow is), the more extreme the associated weather tends to be. For example, heat waves and droughts typically occur under amplified upper-tropospheric ridges, while cold air outbreaks, clouds, and precipitation are frequently associated with various sectors of strong troughs.

In Fig. 5.7, the flow across North America is predominantly meridional, with north-to-south 500-hPa flow located over and just east of the Rockies and southwest-to-northeast flow located west of the ridge off the California coast and just to the east of the primary trough axis in the center of the continent. Zonal flow is observed over New England and eastern Canada, where the geopotential height contours are relatively close together, suggestive of fast wind speeds. Fig. 5.8, a computer analysis of geopotential height and wind speeds for the chart in Fig. 5.4, shows a strong, primarily zonal >180 kt jet streak centered over Ohio and Pennsylvania. Meanwhile, the jet streak over the Rockies is almost entirely meridional, which helps to transport cold air equatorward behind the amplifying trough. Both jet streaks and the equatorward transport of cold air into the developing upper-tropospheric trough were key features in the development of the 1993 Superstorm, as we will see in Chapter 7.

5.3.3   Jet Streaks

Recall that jet streaks are cores of the fastest wind speeds within the jet stream. On the chart in Fig. 5.8, there are three jet streaks, their centers marked with a red "X." Compare the computer analysis to the manually drawn analysis in Fig. 5.5 and also to your own isopleth analysis. As we will see in Chapters 7 and 9,

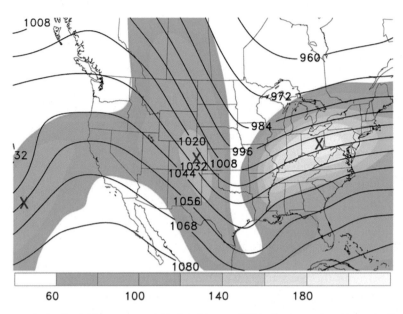

Fig. 5.8 Computer analysis of 250-hPa geopotential height (dam, *solid black contours*) and isotachs (kt, *shaded*) across North America at 00Z 13 March 1993.

jet streaks can play an important role in determining sky conditions, as well as the intensity of surface cyclones and anticyclones. Jet streaks tend to propagate downwind faster than troughs and ridges, and their position relative to a trough or ridge is important (see Chapter 7).

Fig. 5.9 shows a schematic of a straight zonal jet streak, with the center indicated by a red "X." It is important to think of jet streaks as if we are flying through them on an aircraft, with the geostrophic wind to your back. In the case of Fig. 5.9, the wind is westerly, so we will "fly" through the jet streak from left to right. The first (western) half of the jet streak is called the *entrance (rear) region*; here, wind speeds become faster as we move toward the center of the jet streak. The latter (eastern) half of the jet streak is the *exit (front) region*, in which wind speeds decelerate as we move farther away from the center of the jet streak. Now examine the four *jet streak quadrants* labeled in red in Fig. 5.9. The entrance and exit regions are each split into two quadrants, left and right, relative to the wind direction through the jet streak.

Again, examine the strong jet streak centered over Ohio and Pennsylvania in Fig. 5.8. To identify all four quadrants of this streak, we must imagine we are flying through it from southwest to northeast, with the geostrophic wind to our back and low

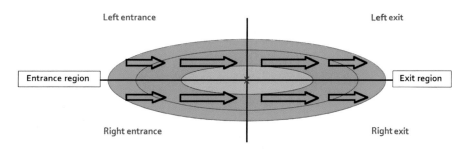

Fig. 5.9 Schematic of a west-east oriented upper-tropospheric jet streak: the *green and blue shadings* are indicative of isotachs, with the largest wind speeds located near the center of the jet streak and marked with a *red* "X." Geostrophic wind (*black arrows*), jet streak regions (*black rectangles*) and quadrants (*red*) are labeled.

geopotential height to our left. As a result, the location of each jet streak quadrant is as follows:

- Right entrance: Tennessee, Mississippi, Alabama.
- Left entrance: Missouri, Illinois.
- Right exit: Eastern North Carolina and the adjacent Atlantic Ocean.
- Left exit: Northern New York, Vermont, southeastern Ontario, southwestern Quebec.

In Chapter 7, we will discuss the implications of each jet streak quadrant on surface features and sensible weather.

5.4 Questions and Practice Exercises

1. Using a blank chart of 500-hPa radiosonde data from your instructor or the most recent chart of North America from NOAA or a similar source, draw geopotential height contours and isotherms using proper plotting conventions.
2. Repeat the exercise in Question 1, but for 250-hPa geopotential height contours and isotachs.
3. For your completed manual analysis in Question 1, identify all 500-hPa trough and ridge axes using the proper symbols.
4. For your completed manual analysis in Question 2, identify all 250-hPa jet streaks.
5. On the 500-hPa chart in Fig. 5.7, determine a) whether the flow in the following locations is zonal or meridional, and b) the approximate direction of the geostrophic (or gradient) wind. Explain your answers.
   - Southern Oregon.
   - Central Saskatchewan.
6. On the 250-hPa chart in Fig. 5.8, is the jet streak located over the Pacific Ocean zonal or meridional? Explain your answer.
7. On the 250-hPa chart in Fig. 5.8, identify all four quadrants in the jet streak centered over Colorado.

6

LOWER-TROPOSPHERIC CHARTS

6.1   Isopleths on Lower-Tropospheric Charts

In this chapter, we discuss isopleths on lower-tropospheric charts and how to use them to identify important meteorological features. Table 6.1 shows isopleth conventions for 700-hPa, 850-hPa, and surface (mean sea-level) charts. 700- and 850-hPa charts typically feature both geopotential height and temperature, with a 3 dam geopotential height contour interval (Table 6.1) and a 5°C isotherm interval. However, 700-hPa charts also usually include relative humidity (RH), specifically regions where RH $\geq$70%, indicative of possible clouds. A good rule of thumb to identify such regions is if dew point depression is $\leq$3°C (Table 6.1).

Surface charts are constant height (not isobaric) charts at 0 m, i.e., mean sea level. Mean sea-level pressure (MSLP) isobars are plotted every 4 hPa (centered on 1000 hPa) and a measure of temperature is also typically plotted: either surface temperature isotherms or 1000–500-hPa thickness, which as we will see in Chapter 8, can be used as a proxy for the mean temperature in the column. Note that there is no meaningful difference between "SLP" (see Chapters 2 and 3) and "MSLP"; however, the meteorological convention is to use the former when discussing station plots and METAR code, and the latter when plotting isobars.

6.1.1   700-hPa Charts

The 700-hPa level can be thought of as a bridge between the upper troposphere and the surface. Upper-tropospheric features such as troughs and ridges are still evident, but 700-hPa charts can also be used to identify regions of saturated air (large RH) favorable for cloud formation. In this section, we will use the 700-hPa radiosonde measurements plotted in Fig. 6.1, at the same time (00Z 13 March 1993) as the charts in Chapter 5. Recall from Chapter 4 that on a standard upper-air station plot, temperature is

Synoptic Analysis and Forecasting. https://doi.org/10.1016/B978-0-12-809247-7.00006-5

Table 6.1 Standard Isopleth Conventions for Lower-Tropospheric Charts

Chart Type	Geopotential Height	Temperature	Other
700 hPa	Every 3 dam, centered on 300 dam	Every 5°C, centered on 0°C	Dew point depression shaded at ≤3°C (~70% relative humidity) and ≤1°C (~90% relative humidity)
850 hPa	Every 3 dam, centered on 150 dam	Every 5°C, centered on 0°C	None
Surface/ mean sea level	None	United States: every 10°F, centered on 50°F Elsewhere: every 5°C, centered on 10°C Or: 1000–500-hPa thickness every 6 dam, as a proxy for temperature	Mean sea-level pressure (MSLP) every 4 hPa, centered on 1000 hPa

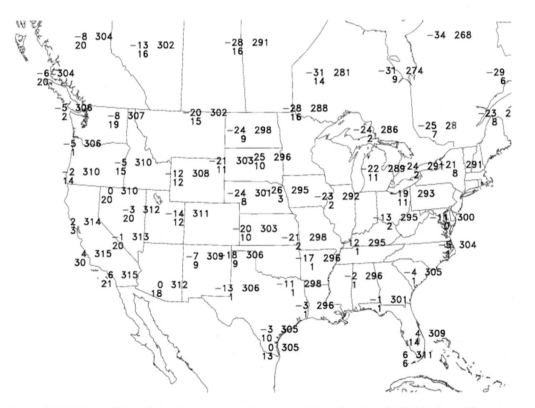

Fig. 6.1 At 00Z 13 March 1993, radiosonde-measured values across North America of: 700-hPa (top left) temperature (°C), (bottom left) dew point depression (°C), and (top right) geopotential height (dam). In order to enhance readability, wind information is not plotted.

plotted at top left, dew point depression at bottom left, and geopotential height at top right (Fig. 6.1).

In order to isopleth Fig. 6.1, we again need to identify the smallest and largest isohypse values. In Fig. 6.1, the smallest geopotential height values are located in northern Quebec, with a minimum station value of 268 dam. Thus the smallest isohypse on the chart will be 270 dam. As on our 500-hPa chart (Chapter 5), the largest geopotential height values are located in southern California, with a maximum station value of 315 dam (Fig. 6.1). Because there are two stations reporting 315 dam, the 315 dam isohypse should be our largest contour value.

In terms of temperature, the coldest air is draped across northern Ontario and Quebec, with a minimum value of $-34°C$, meaning that the minimum isotherm value on the chart will be $-30°C$ (Table 6.1). The warmest 700-hPa air is located in southern California and southern Florida, indicating that we should draw two separate $5°C$ isotherms. Isohypses are plotted using solid contours and isotherms using dashed lines.

Fig. 6.2A shows the complete set of 700-hPa isohypses. As expected, the smallest values are in the east, associated with two 700-hPa trough axes, one across the Great Lakes, and the other along the central Gulf Coast. Meanwhile, a 700-hPa ridge is located across the west, most prominently in southern California, Nevada, and Idaho. In Fig. 6.2B, the coldest air is largely colocated with the trough across the northern Great Lakes, while the warmest air is in southern California and southern Florida. Note that there is a strong temperature gradient oriented from southeast to northwest ahead of the Gulf Coast trough. This temperature gradient serves as "fuel" for the developing lower-tropospheric cyclone and precipitation, discussed further in Chapter 8.

Fig. 6.2C shades regions of RH $\geq70\%$ in green; this is done by shading all stations with a dew point depression $\leq3°C$ and subsequently any area between such stations. The majority of the large RH region is located at or just ahead of the 700-hPa trough in the eastern United States. As we will see later in this chapter, knowing the position of 700-hPa troughs and ridges can be useful for diagnosing regions of potential clouds.

6.1.2   850-hPa Charts

We now move to 850-hPa charts, which can be thought of as similar to surface charts. Recall that 850 hPa is on average 1500 m (150 dam) above sea level, which in regions of high terrain (e.g., Rocky Mountains) can be located below the surface. Therefore it is important to take 850-hPa features with a grain of salt over major mountain ranges such as the Rockies or Alps. In these

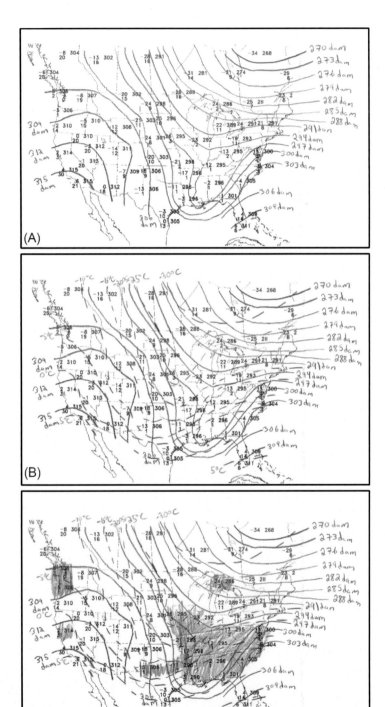

Fig. 6.2 For the chart in Fig. 6.1, manually drawn isohypses (geopotential height contours) every 3 dam (*solid black*), isotherms every 5°C (*dashed red*), and dew point depression ≤3°C (*green shaded*):
(A) isohypses only,
(B) isohypses and isotherms, and (C) the complete chart.

regions, 700-hPa charts can be used as a reasonable substitute. In this section, we will use the chart of 850-hPa radiosonde measurements across North America in Fig. 6.3, which is at the same time (00Z 13 March 1993) as Fig. 6.1 and the charts in Chapter 5.

To isopleth Fig. 6.3, we again need to diagnose the smallest and largest isohypse values, which are 132 and 159 dam, respectively (Table 6.1). The smallest geopotential height values are once again located in northern Quebec, but on this chart the largest geopotential height values are across Wyoming and northern Colorado. Because there are two stations reporting 159 dam, we can use the 159 dam isohypse as our largest contour value. In Fig. 6.4A, several closed geopotential height contours are evident across the US Rockies and Gulf Coast. Unlike the mid- and upper-troposphere, where most features are open waves, lower-tropospheric high and low pressures are commonly "closed" systems, featuring at least one closed isohypse (isobar). Because the feature over the US Rockies is a geopotential height maximum, it is an anticyclone, while the feature along the Gulf Coast is a

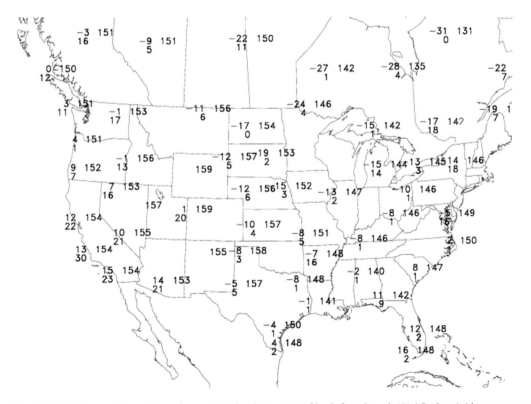

Fig. 6.3 At 00Z 13 March 1993, radiosonde-measured values across North America of: 850-hPa (top left) temperature (°C), (bottom left) dew point depression (°C), and (top right) geopotential height (dam). In order to enhance readability, wind information is not plotted.

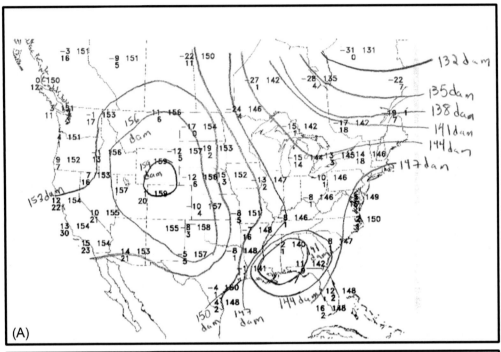

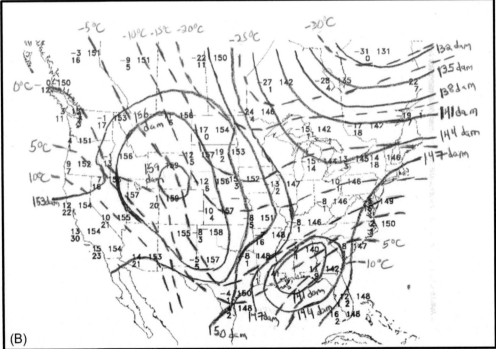

Fig. 6.4 For the chart in Fig. 6.3, manually drawn isohypses (geopotential height contours) every 3 dam (*solid black*) and isotherms every 5°C (*dashed red*): (A) isohypses only and (B) isohypses and isotherms.

cyclone. We examine these features more closely in the next section.

For temperature, the coldest air at 850 hPa is still in northern Ontario and Quebec, with a minimum isotherm value of $-30°C$ (Table 6.1). Similar to at 700 hPa, the warmest 850-hPa air is in California and Florida, indicating the need for two disparate 10°C isotherms. As at 700 hPa (Fig. 6.2), isohypses should be plotted using solid contours and isotherms using dashed lines. In Fig. 6.4A, notice areas such as northwestern Ontario or the southeastern United States where the geopotential height contours are largely perpendicular to the isotherms. It is useful for the analyst or forecaster to train his/her eye to identify such regions, as they represent important physical mechanisms that affect weather forecasts, discussed in Chapter 8.

6.1.3 MSLP Charts

Recall that MSLP is the pressure a location would have if it were located at mean sea level (0 m in elevation). METAR observations of surface pressure are converted into MSLP by the ASOS or observer (Chapters 2 and 3). Because pressure decreases drastically with increasing altitude, MSLP allows us to determine horizontal variations in pressure without considering the impact of altitude. As such, we are able to diagnose surface high and low pressure systems, which have a large impact on the weather that we experience. Fig. 6.5 shows surface station plots at 06Z 13 March 1993; this chart is taken from 6 h later than the 1993 Superstorm charts that we have discussed, in order to better observe the development of the intense surface cyclone. To facilitate drawing isobars, only MSLP and wind barbs are plotted in Fig. 6.5.

MSLP isobars are drawn every 4 hPa, centered on 1000 hPa (Table 6.1). In Fig. 6.5, we see that 991.2 hPa in southern Georgia is the lowest MSLP value on the chart, while 1034.9 hPa in Denver and Amarillo, Texas is the highest MSLP value. Therefore our lowest isobar will be 992 hPa and our highest will be 1032 hPa. It is useful to start with either the lowest or highest isobar value and work progressively from there. Remember that surface cyclones and anticyclones are frequently closed systems, so there is often more than one isobar per value on a single analysis chart.

Fig. 6.6A shows the 992-hPa isobar circled around the only MSLP station value (991.2 hPa) that is <992 hPa; this observation is near the center of the surface cyclone, as also evidenced by the cyclonic (counterclockwise in the Northern Hemisphere) winds around the 992-hPa isobar. For a surface low pressure (cyclone), MSLP values less than a particular isobar should be inside of that

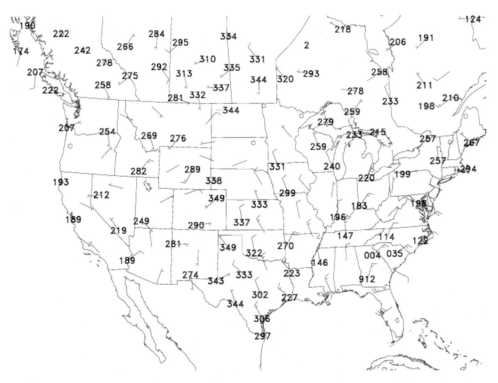

Fig. 6.5 At 06Z 13 March 1993, selected surface station values of MSLP (hPa) and wind (knots, barbs). MSLP is plotted using the surface station plot conventions discussed in Chapter 2. In order to enhance readability, temperature, cloud cover, and present weather information are not plotted.

isobar, while MSLP values greater than that isobar's value should be located on the outside of the isobar. In contrast, MSLP values at the center of a surface high pressure (anticyclone) should be greater than the isobar surrounding them (Fig. 6.6B). We observe this for the anticyclone located over much of the U.S. Great Plains and Canadian Prairies.

On a surface analysis over a large area, few MSLP isobars will typically traverse the entire map, especially in a case with amplified pressure systems such as those in Fig. 6.6C. As a result, drawing, e.g., 500-hPa isohypes can be a very different experience from drawing MSLP isobars. More than one and occasionally more than two isobars of the same value are required on a single chart. We see this in Fig. 6.6C with the 1020-hPa isobar value, for which there are four total contours: one in British Columbia, a second in Oregon and California, a third in the eastern United States, and a fourth in Quebec.

In general, winds rotate anticyclonically around a surface high pressure and cyclonically around a surface low pressure,

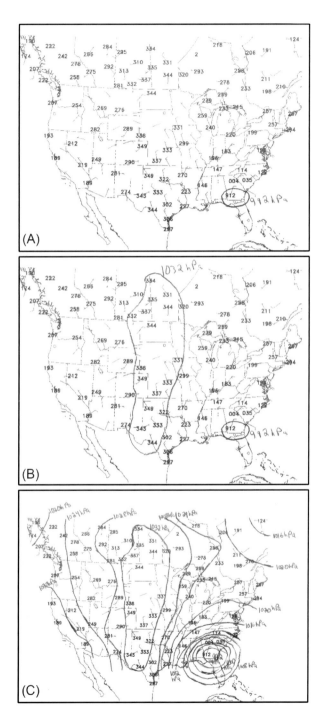

Fig. 6.6 For the chart in Fig. 6.5, manually drawn MSLP isobars every 4 hPa: (A) the smallest MSLP value (992 hPa) on the chart, (B) the smallest and largest (1032 hPa) MSLP value on the chart, and (C) the complete chart.

although boundary-layer friction will typically deflect the wind approximately 30°–45° toward low pressure and away from high pressure. This relationship between the surface wind and MSLP isobars is a version of Buys Ballot's law (Chapter 5), modified to include the effects of friction. Observe the surface wind directions near the centers of the surface anticyclone and cyclone in Fig. 6.6. Near the cyclone, the wind is blowing counterclockwise around the cyclone, but at an angle toward low pressure; near the anticyclone, the wind is blowing clockwise around the 1032-hPa isobar, but generally at an angle away from high pressure.

6.2 Feature Identification

6.2.1 700-hPa Troughs and Ridges

Fig. 6.7 shows a computer-analyzed 700-hPa chart that includes isohypses, isotherms, and regions of RH ≥70%. Compare it to your own 700-hPa chart; they should look similar! As we saw in Chapter 5, the smallest geopotential heights are located in the east and the largest in the west. Example 700-hPa trough and ridge axes are marked in Fig. 6.7 to facilitate discussion. As we found in our own analysis (Fig. 6.2), the majority of the large RH region in the east is located at or ahead (downwind) of the trough axis. In the west, large RH values are located upstream (west of) the 700-hPa ridge axis. Due to the physical processes discussed in

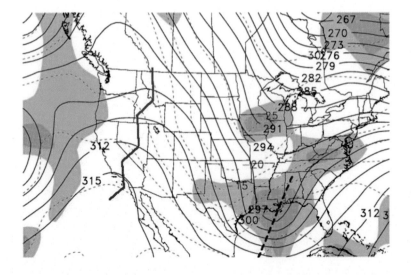

Fig. 6.7 Computer analysis of 700-hPa geopotential height (dam, *solid black contours*) and temperature (°C, *dashed red contours*) across North America at 00Z 13 March 1993. A 700-hPa trough (*dashed blue line*) and ridge (*jagged red line*) are marked using standard symbols.

Chapters 7 and 8, the position of 700-hPa trough and ridge axes can be used as rough estimates of possible cloudy regions.

A good forecast rule of thumb is that after a 700-hPa trough axis passes, RH will begin to drop and skies will start to clear. In contrast, after a 700-hPa ridge axis moves through a point, RH will start to increase and cloud cover will become more prominent. We observe this for most of the troughs and ridges in Fig. 6.7, where from left to right, large RH values are predominantly located ahead of trough axes, but behind ridge axes.

6.2.2 Cyclones and Anticyclones

On the 500-hPa charts in Chapter 5, we learned how to identify upper-tropospheric troughs and ridges. Here, we will use 850-hPa and MSLP charts to spot surface cyclones and anticyclones. Fig. 6.8 shows a computer analysis of 850-hPa geopotential height and temperature. Analogous to an upper-tropospheric trough, a lower-tropospheric cyclone is a geopotential height minimum, while a lower-tropospheric anticyclone is a geopotential height maximum and is analogous to an upper-tropospheric ridge. The convention on lower-tropospheric charts is to mark cyclones with a red "L" and anticyclones with a blue "H." The cyclone over the Gulf Coast and the anticyclone in southern Wyoming in Fig. 6.8 should closely match our manual analysis in Fig. 6.4.

The 850-hPa chart in Fig. 6.8 also shows that the isohypses and isotherms are offset from each other. This is not unusual; the

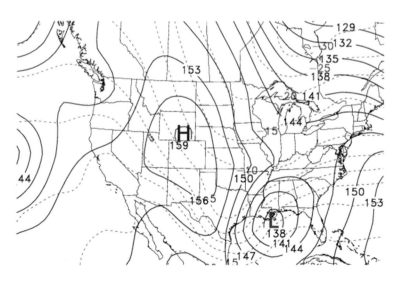

Fig. 6.8 Computer analysis of 850-hPa geopotential height (dam, *solid black contours*) and temperature (°C, *dashed red contours*) across North America at 00Z 13 March 1993. 850-hPa cyclones and anticyclones are marked with a *red* "L" and *blue* "H," respectively.

coldest temperatures are often located upstream of a surface cyclone, particularly developing or intensifying cyclones, while the warmest temperatures may be found upstream of a surface anticyclone. In Fig. 6.8, an axis of minimum 850-hPa temperatures is located over southeastern Texas and the western Gulf of Mexico, while the surface cyclone is centered farther east. Similarly, the axis of the warmest temperatures is found over the west coast of North America, while the surface anticyclone is located over the Rockies. This results in regions where the isohypses and isotherms are nearly perpendicular to each other, such as over Alabama, Georgia, and Florida; in Chapters 8 and 9, we will discuss the physical importance of these regions.

Finally, Fig. 6.9 shows a computer-analyzed MSLP chart at 06Z 13 March 1993. There are some slight differences between this chart and our manual analysis (Fig. 6.6C), including a higher minimum MSLP for the developing surface cyclone. The intense nature of the surface cyclone and anticyclone are exemplified by the strong MSLP gradient between them; just as the geopotential height gradient is proportional to wind speed at upper levels, the MSLP gradient is here as well. Thus we can conclude that relatively strong winds are present from eastern Texas into Alabama, Georgia, and the Carolinas. In contrast, the lack of a MSLP gradient across southern California (Fig. 6.9) is indicative of weak surface winds in that region.

It is also useful to compare the location of the surface cyclone and anticyclone in Fig. 6.9 to the corresponding 850-hPa features 6 h earlier (Fig. 6.8). Typically, 850-hPa and surface cyclones and

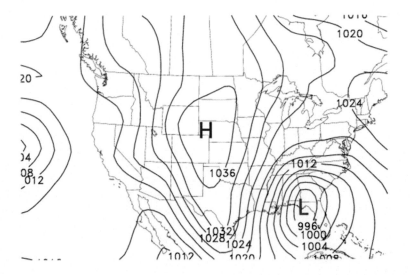

Fig. 6.9 Computer analysis of MSLP (hPa, *solid black contours*) across North America at 06Z 13 March 1993. Surface cyclones and anticyclones are marked with a *red* "L" and *blue* "H," respectively.

anticyclones will be located near or on top of one another. As we can see, the cyclone has moved from near New Orleans at 00Z (Fig. 6.8) to the Florida Panhandle at 0600Z (Fig. 6.9). Meanwhile, the anticyclone has propagated slightly eastward as well. In Chapters 8 and 9, we will discuss how to use lower-tropospheric charts to forecast the motion of lower-tropospheric weather systems.

6.3 Questions and Practice Exercises

1. Using a blank chart of 700-hPa radiosonde data from your instructor or the most recent chart of North America from NOAA or a similar source, draw geopotential height contours and isotherms using proper plotting conventions. Shade areas of RH $\geq 70\%$ in green.
2. Repeat the exercise in Question 1, but for 850-hPa geopotential height contours and isotherms.
3. For your completed manual analysis in Question 1, identify all 700-hPa trough and ridge axes using the proper symbols.
4. For your completed manual analysis in Question 2, identify all 850-hPa cyclones and anticyclones using the proper symbols.
5. What is the typical geopotential height contour interval used on lower tropospheric (e.g., 850-hPa) charts?
6. Based on standard plotting conventions, should you ever draw a 1006-hPa isobar on a MSLP analysis chart? Explain why or why not.
7. Clouds are more likely (ahead/behind) of a 700-hPa ridge axis.

UPPER-TROPOSPHERIC PROCESSES

7.1   Divergence and Vertical Motion

Prior to discussing upper-tropospheric processes and how they relate to weather forecasting, it is useful to review why air rises or sinks. Recall that *divergence* is defined as air spreading apart and/or rushing away, while *convergence* means air coming together and/or slowing down. Horizontal mass divergence and convergence can be related to vertical air motion through the *mass continuity principle*[1] *(continuity equation)*, which relates a change in mass of a volume of air to the net flow of mass into that volume.

In the troposphere, ascent (rising vertical motion) is caused by surface convergence, upper-tropospheric divergence, or a combination of the two (Fig. 7.1A). When air converges at the surface, it cannot travel into the ground. Therefore to preserve mass continuity, the air must rise. At the tropopause, air cannot continue into the stratosphere because that layer is extremely stable and resists vertical motion; thus rising air must diverge at the tropopause. Through the mass continuity principle, we can associate surface convergence and upper-tropospheric divergence with ascent, which, in turn, is associated with clouds and precipitation.

In contrast, descent (sinking vertical motion) is caused by surface divergence, upper-tropospheric convergence, or a combination of the two (Fig. 7.1B). When air converges at the tropopause, it cannot move up into the stratosphere; therefore it must descend toward the surface. At the surface, the sinking air cannot go into the ground, so to preserve mass continuity, it will diverge. Thus we can say that surface divergence and upper-tropospheric

[1] American Meteorological Society, 2012: Glossary of Meteorology, available online at: http://glossary.ametsoc.org/wiki/Equation_of_continuity.

Synoptic Analysis and Forecasting. https://doi.org/10.1016/B978-0-12-809247-7.00007-7

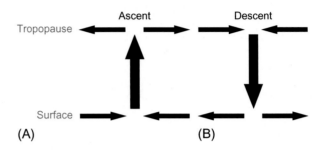

Fig. 7.1 Schematics of air motion at the surface and tropopause for (A) ascent (rising vertical motion) and (B) descent (sinking vertical motion). Horizontal arrows pointing away from (towards) each other indicate divergence (convergence).

convergence are both associated with descent, which, in turn, results in less cloud cover and no precipitation.

In this chapter, we will focus on the upper-tropospheric portion of the three-dimensional circulation pictured in Fig. 7.1 and its applicability to weather forecasting. Chapter 8 discusses the surface aspects, while Chapter 9 ties everything together using a complete three-dimensional framework.

7.2 Jet Streaks

7.2.1 The Four-Quadrant Model

In Chapter 5, we reviewed geostrophic balance and Buys Ballot's law, concluding that in the Northern Hemisphere (NH) with the geostrophic (or gradient) wind at your back, low pressure (geopotential height) is on your left. We used this principle to introduce the straight jet streak four-quadrant model, a schematic of which is shown again in Fig. 7.2. Here, we will associate each jet streak quadrant with upper-level divergence or convergence, and thus vertical motion.

Examine the jet entrance region in Fig. 7.2; as the wind blows toward the jet streak core, it accelerates. When the wind accelerates, geostrophic balance is briefly disturbed, such that the magnitude of the pressure gradient force (PGF) slightly exceeds that of the Coriolis Force. Because the PGF always points toward low pressure, PGF exceeding Coriolis means that a component of the wind in the entrance region is directed toward lower pressure (geopotential height), as shown by the orange arrow in Fig. 7.2. In the exit region, the wind decelerates as it moves away from the jet core. Geostrophic balance is again briefly disrupted in the exit region, except here PGF is smaller than Coriolis and so a

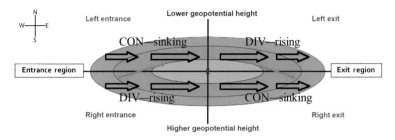

Fig. 7.2 Schematic of a Northern Hemisphere (NH) west-east oriented upper-tropospheric jet streak: the *green and blue shadings* are indicative of isotachs, with the largest wind speeds located near the center of the jet streak and marked with a *red* "X." Typical of the NH, lower geopotential height (pressure) is located to the north and higher geopotential height (pressure) to the south. Geostrophic wind (*black arrows*), jet streak regions (*black rectangles*), and quadrants (*red*) are labeled; the *orange arrows* represent the deflection of air motion in each jet region, and the *black text* in each quadrant defines whether that quadrant features upper-level divergence or convergence, as well as the resulting vertical motion.

component of the wind (Fig. 7.2) blows away from low pressure (geopotential height), toward high pressure (geopotential height).

Now look at the two orange arrows in Fig. 7.2 and observe whether each arrow converges or diverges with the westerly geostrophic wind (black arrows) in each jet streak quadrant. In the entrance region, the right entrance features upper-level divergence, associated with ascent, while the left entrance is marked by upper-level convergence, associated with descent. In the exit region, the right exit is marked by upper-level convergence (descent), while the left exit features upper-level divergence (ascent). Using this four-quadrant model of straight jet streaks, we can conclude that in the NH, the right entrance and left exit of a jet streak are most favorable for ascent, clouds, and precipitation. In contrast, the left entrance and right exit of a NH jet streak are areas promoting descent, and thus clouds and precipitation.

As we will see over the next three chapters, jet streak divergence is only one of three synoptic-scale mechanisms associated with vertical motion. However, it is important for the analyst and forecaster to be able to recognize these regions when looking at a jet stream chart. Fig. 7.3 shows the 250-hPa chart at 00Z 13 March 1993, during the development of the 1993 Superstorm (Chapter 5). On this chart, the two straight jet streaks over the United States are divided into quadrants. Remember to examine each jet streak as if you are flying through it with the geostrophic wind at your back. Regions of upper-tropospheric convergence are labeled with a

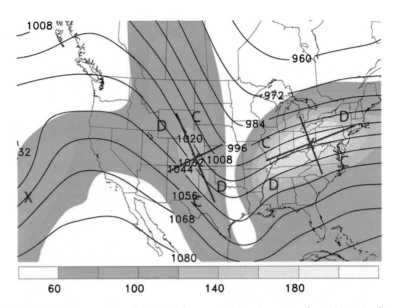

Fig. 7.3 250-hPa geopotential height (dam, *solid black contours*) and isotachs (kt, *shaded*) across North America at 00Z 13 March 1993. The two jet streaks located over the United States are divided into regions and quadrants (*red lines*), with upper-level divergence (*red "D"*) and convergence (*red "C"*) specified in each quadrant.

"C," while regions of upper-tropospheric divergence are labeled with a "D." For the easternmost jet streak, the right entrance quadrant is located across Mississippi, Alabama, and Tennessee, while the left exit is found in northern New York and northern New England. These two quadrants are where, using jet streak analysis alone, we can expect upper-tropospheric divergence and ascent. In contrast, the adjacent left entrance and right exit quadrants drape across Illinois/Indiana and the coastal Mid-Atlantic, respectively; these regions will experience upper-level convergence, and thus descent.

7.2.2 Straight Versus Curved Jet Streaks

To this point, we have discussed regions of straight jet streaks and their association with vertical motion. Curved jet streaks (i.e., jet streaks around the base of a trough or apex of a ridge) should not be interpreted using the four-quadrant model. To the first order, curved jet streaks should instead be analyzed in terms of entrance and exit regions only.[2] In a cyclonically curved jet streak (i.e., a jet streak in the base of a trough), the entire entrance region is marked by upper-tropospheric convergence, while the entire exit region features upper-tropospheric divergence. For an anticyclonically curved jet streak (i.e., a jet streak through the apex of a ridge), it is the reverse: the entire entrance region is divergent, while the exit region is convergent. We can therefore conclude that the exit region of a cyclonically curved jet streak and the entrance region of an anticyclonically curved jet streak are regions of ascent. It is important that the analyst or forecaster recognize the difference between straight and curved jet streaks before choosing which divergence model to use.

Fig. 7.4 shows a 250-hPa chart at 18Z 13 March 1993, during the latter part of Superstorm 1993. Two jet streaks are highlighted: a straight west-east jet streak over eastern Canada and a curved jet streak over the Gulf Stream within the base of the negatively tilted upper-tropospheric trough. The straight jet streak to the north follows the four-quadrant model, indicating that the right entrance is divergent, while the left entrance is convergent. However, the southern jet streak follows the two-region model, indicating that the entire exit region near the Mid-Atlantic coast is divergent and the entrance region west of the trough axis is convergent. Note that the right entrance region of the northern jet streak and the exit region of the curved southern jet streak are in close proximity; this is called a *dual jet divergence* region. Dual jet divergence regions tend to be present in cases of rapid surface cyclogenesis such as the 1993 Superstorm. They typically feature very strong upper-tropospheric divergence, which in turn causes strong ascent.

7.3 Vorticity

7.3.1 Definition

Divergence and convergence are the physical mechanisms that cause vertical motion. However, they tend to be very "noisy" fields that are notoriously difficult to accurately measure and plot on an isobaric chart. Therefore meteorologists frequently use a proxy for upper-tropospheric divergence and convergence called *vorticity*,[3] which is a vector measure of the rotation (spin) of an air

[2] Moore, J. T., and G. E. Vanknowe, 1992: The effect of jet-streak curvature on kinematic fields. *Mon. Wea. Rev.,* **120,** 2429–2441.
[3] American Meteorological Society, 2012: Glossary of Meteorology, available online at: http://glossary.ametsoc.org/wiki/Vorticity.

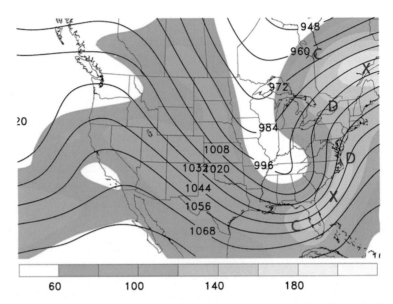

Fig. 7.4 250-hPa geopotential height (dam, *solid black contours*) and isotachs (kt, *shaded*) across North America at 18Z 13 March 1993. Divergent ("D") and convergent ("C") regions are indicated in *red* for the two jet streaks (*one curved, one straight*) located in eastern North America.

parcel. There are two components of total, or *absolute vorticity*[4]: (1) *relative vorticity*[5] and (2) *planetary vorticity*. Relative vorticity is the rate of spin of an air parcel for a fixed coordinate system on the Earth; in other words, it is the spin associated with weather systems, i.e., cyclones and anticyclones. Planetary vorticity is purely due to the rotation of the Earth. Looking down upon the North Pole, Earth spins counterclockwise; if we look down onto the South Pole, Earth rotates clockwise. Therefore planetary vorticity is always counterclockwise in the NH and clockwise in the Southern Hemisphere (SH).

To calculate absolute vorticity, we need to add relative vorticity and planetary vorticity. In the NH, air rotates counterclockwise around a cyclone, which is defined as cyclonic relative vorticity. In the SH, air rotates clockwise around a cyclone; this is also called cyclonic relative vorticity. In this text, we primarily use the terms cyclonic and anticyclonic for relative vorticity because they are hemisphere independent; alternative terms are positive and

[4]American Meteorological Society, 2012: Glossary of Meteorology, available online at: http://glossary.ametsoc.org/wiki/Absolute_vorticity.
[5]American Meteorological Society, 2012: Glossary of Meteorology, available online at: http://glossary.ametsoc.org/wiki/Relative_vorticity.

negative relative vorticity, which mean different things in each hemisphere. In the NH, cyclonic relative vorticity is positive, but in the SH cyclonic relative vorticity is negative. Along the same lines, anticyclonic relative vorticity is negative in the NH, but positive in the SH.

Because air around cyclones rotates in the same direction as the Earth in each respective hemisphere, absolute vorticity always has a larger absolute value in cyclones than in anticyclones, regardless of hemisphere. Anticyclones, which rotate clockwise in the NH and counterclockwise in the SH, typically exhibit small absolute vorticity. This is because although anticyclonic relative vorticity (negative in the NH) is of opposite sign as planetary vorticity (positive in the NH), the magnitude of planetary vorticity is almost always larger than that of relative vorticity. For example, assume the following values in the NH midlatitudes:

- Planetary vorticity $= 10.3 \times 10^{-5}\,\text{s}^{-1}$.
- Relative vorticity of a NH 500-hPa cyclone $= 16 \times 10^{-5}\,\text{s}^{-1}$.
- Relative vorticity of a NH 500-hPa anticyclone $= -6 \times 10^{-5}\,\text{s}^{-1}$.

To calculate the respective absolute vorticities of the 500-hPa cyclone and anticyclone, we need to add planetary and relative vorticity for each. In the cyclone case, the absolute vorticity is $26.3 \times 10^{-5}\,\text{s}^{-1}$. In the anticyclone case, the absolute vorticity is $4.3 \times 10^{-5}\,\text{s}^{-1}$. As we can see from this simplistic example, NH cyclones and anticyclones in the midlatitudes both have positive absolute vorticities. However, the magnitude of absolute vorticity of the cyclone is much larger than that of the anticyclone.

7.3.2 Causes of Relative Vorticity

In the last section, we saw that large absolute vorticity is indicative of cyclones, while smaller values are associated with anticyclones. Here, we introduce the two mechanisms that cause relative vorticity: *curvature* and *shear*. Relative vorticity due to curvature is the spin of an air parcel due to the curvature of its path, such as in troughs and ridges (Fig. 7.5A). As an air parcel moves into the base of a trough, it will begin to spin cyclonically (counterclockwise in the NH), acquiring cyclonic relative vorticity. An air parcel moving into a ridge axis will spin anticyclonically (clockwise in the NH), resulting in anticyclonic relative vorticity (Fig. 7.5A). Therefore troughs are regions of cyclonic relative vorticity (large absolute vorticity), while ridges feature anticyclonic relative vorticity (small absolute vorticity).

The other mechanism that can result in air parcels acquiring relative vorticity is horizontal wind shear. In Fig. 7.5B, two

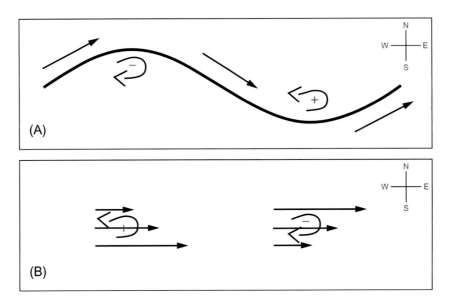

Fig. 7.5 Schematics of (A) curvature and (B) shear vorticity. Wind speed and direction is specified by the *black arrows*. The *blue* (*red*) positive (negative) sign indicates cyclonic (anticyclonic) vorticity, which is positive (negative) in the NH.

scenarios are presented: on the left, geostrophic wind speed increases toward the south, while on the right, it increases toward the north. Now insert an imaginary pinwheel in the middle of each scenario; the pinwheel on the left will spin counterclockwise, while the one on the right will spin clockwise. Therefore decreasing westerly geostrophic wind with increasing latitude results in cyclonic relative vorticity due to shear, while increasing westerlies with increasing latitude cause anticyclonic relative vorticity; this principle can be applied in both the NH and SH.

In the mid- and upper-troposphere, large absolute vorticity in a trough is caused by a combination of curvature and shear, and the same is true for small absolute vorticity in a ridge. As a result, the maximum cyclonic relative vorticity is typically found on the poleward side of a trough, where both curvature and shear promote cyclonic relative vorticity. For a ridge, the maximum anticyclonic relative vorticity (and thus minimum absolute vorticity) is usually located on the equatorward side, where both curvature and shear promote anticyclonic rotation.

7.3.3   Identification

Fig. 7.6A shows 500-hPa geopotential height and geostrophic absolute vorticity at 12Z 13 March 1993, during the rapid cyclogenesis of the Superstorm. In this text, all relative and absolute vorticity calculations are performed using the geostrophic wind, because we

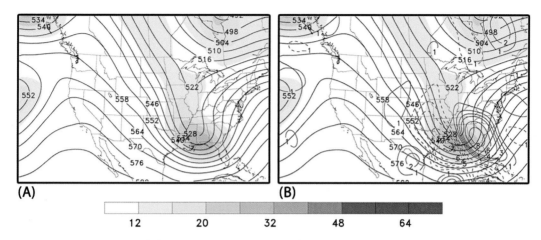

Fig. 7.6 At 12Z 13 March 1993: (A) 500-hPa geopotential height (dam, *solid black contours*) and geostrophic absolute vorticity ($\times 10^{-5}$ s^{-1}, *shaded*), and (B) 500-hPa geopotential height (dam, *solid black contours*) and geostrophic absolute vorticity advection ($\times 10^{-9}$ s^{-2}, *blue contours*, solid for CVA, *dashed* for AVA) added.

only use vorticity above the friction (boundary) layer. The largest absolute vorticity values on the 500-hPa chart are located on the northern side of the base of the strong 500-hPa trough centered across Alabama and Mississippi. This is where cyclonic relative vorticity is largest, promoted by a combination of curvature and shear. Similarly, large absolute vorticity values are observed near the trough across Hudson/James Bay, on the northern fringe of the chart.

The primary ridge axis in Fig. 7.6A is located along the US west coast. Here, absolute vorticity values are small, but still positive, because even though relative vorticity is anticyclonic (negative in the NH), the magnitude of planetary vorticity is larger. Nevertheless, on an absolute vorticity chart, ridges are marked by small values. The final area of interest is the yellow shading stretching across central Saskatchewan, Manitoba, and North Dakota. The isohypses in this region are straight, indicating a lack of relative vorticity due to curvature. Thus we can infer that the moderately large values of absolute vorticity are representative of cyclonic relative vorticity due to shear (Fig. 7.6A). Cyclonic relative vorticity due solely to shear will frequently appear as an elongated filament of absolute vorticity, as seen in Fig. 7.6A.

7.4 Vorticity Advection

7.4.1 Definition

Vorticity is a useful proxy for upper-tropospheric divergence and convergence. In reality, however, it is *vorticity advection* that is most closely associated with these processes. Recall that *advection* is the transport of any meteorological quantity (e.g., temperature, moisture, vorticity) by the wind. Here, we will be discussing the advection of vorticity by the geostrophic wind on 500-hPa charts. The key to identifying regions of vorticity advection is to ask yourself "to where is the geostrophic wind transporting vorticity?"

Fig. 7.7 shows a schematic of a 500-hPa ridge-trough pattern in the NH. Typical of the NH midlatitudes, lower geopotential height is located to the north, indicative of a westerly geostrophic wind. Because the ridge axis is the location of maximum anticyclonic relative vorticity (minimum absolute vorticity), the region just downstream of the ridge is located in *Anticyclonic Vorticity Advection* (AVA). Meanwhile, a trough axis is colocated with maximum cyclonic relative vorticity (maximum absolute vorticity), so the region downstream of the trough is characterized by *Cyclonic Vorticity Advection* (CVA). As indicated in Table 7.1, AVA is associated with upper-tropospheric convergence and thus descent, while CVA is associated with upper-tropospheric divergence and thus ascent.

500-hPa vorticity advection is the second of three synoptic-scale forcing mechanisms associated with vertical motion, the first of which was jet streak divergence/convergence, and the third of which (lower-tropospheric temperature advection) we introduce in Chapter 8. It is important to recognize that in order to diagnose the sense of vertical motion, all three mechanisms should be evaluated. In other words, 500-hPa vorticity advection should not replace jet streak divergence/convergence and vice versa.

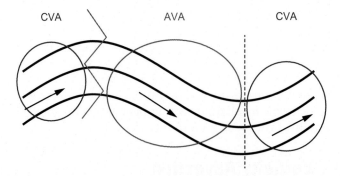

Fig. 7.7 Schematic of a 500-hPa ridge-trough pattern. The ridge (trough) axis is marked by the red jagged (blue dashed) line. Regions of cyclonic vorticity advection (CVA) are indicated in *blue* and regions of anticyclonic vorticity advection (AVA) are indicated in *red*.

Table 7.1 Summary of 500-hPa Vorticity Advection and Its Association With Divergence/Convergence and Vertical Motion

	CVA	AVA
Location	Downstream of a trough	Downstream of a ridge
Upper-tropospheric divergence/convergence	Divergence	Convergence
Vertical motion	Ascent	Descent

7.4.2   Identification

Examine Fig. 7.6B, which shows the same chart as in Fig. 7.6A, but with CVA (solid blue contours) and AVA (dashed blue contours) added. Not surprisingly, the strongest CVA on the chart is centered over Georgia, just downstream of the 500-hPa trough axis (Fig. 7.7). The largest AVA is located just upstream of the trough axis, across Louisiana and Arkansas. For the weaker trough over Hudson Bay, there is CVA present on the eastern side, but it is weaker than the CVA in the southeast United States. Strong CVA (AVA) is indicative of strong upper-tropospheric divergence (convergence), which in turns promotes strong ascent (descent). If CVA and jet-streak divergence are colocated, the chance of heavy precipitation increases, because two synoptic-scale mechanisms are promoting ascent.

Finally, we use Fig. 7.8 to introduce 500-hPa *closed* and *cutoff lows*. A 500-hPa *closed low* is defined as any 500-hPa cyclone (trough) that has at least one closed geopotential height contour. A *cutoff* low is a 500-hPa closed low that is entirely removed from the westerly polar jet stream and usually contains more than one closed isohypse. In Fig. 7.8, the polar jet stream (strongest geopotential height gradient) is located though central Canada, far to the north of the two cyclones over the United States. We can therefore stipulate that both the eastern and western US closed lows are cutoff lows. Absolute vorticity around closed and cutoff lows tends to be relatively disperse (Fig. 7.8), especially compared to a strong open-wave trough such as the 1993 Superstorm (Fig. 7.6).

For the western US cutoff low (Fig. 7.8), there is only one closed geopotential height contour, meaning that the geostrophic wind still promotes CVA downwind (east) of the closed low center. The system in the eastern United States has multiple closed

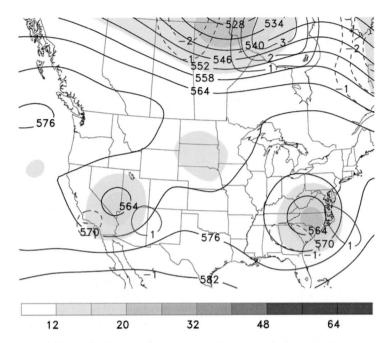

Fig. 7.8 At 00Z 8 May 2013: 500-hPa geopotential height (dam, *solid black contours*), geostrophic absolute vorticity ($\times 10^{-5}$ s^{-1}, *shaded*), and geostrophic absolute vorticity advection ($\times 10^{-9}$ s^{-2}, *blue contours*, solid for CVA, *dashed* for AVA).

isohypses; the geostrophic wind, always parallel to the isohypses, blows counterclockwise around the cutoff low. Because large values of absolute vorticity are scattered around the cutoff low, vorticity is essentially advected in a circle around the entire system. To this point, in Fig. 7.8, weak CVA is located on the north side of the system, while weak AVA is on the southwest side. In general, CVA and AVA around closed and cutoff cyclones tend to be of lesser magnitude than CVA and AVA associated with open-wave troughs and ridges. This results in scattered clouds and precipitation, but not widespread or continuous coverage, as would be found, e.g., over Georgia in Fig. 7.6B. In Chapter 9, we will discuss how we can use vorticity advection to predict the motion of 500-hPa systems, including closed and cutoff lows.

7.5 Questions and Practice Exercises

1. Using only your knowledge of jet streaks, describe the direction of vertical motion and impact on clouds and precipitation over Rapid City, South Dakota in Fig. 7.3.
2. The exit region of an anticyclonically curved jet streak is (convergent/divergent), promoting (ascent/descent).
3. Name the two causes of relative vorticity.
4. Why is absolute vorticity in the NH almost always positive?
5. You are trying to calculate the absolute vorticity of a midlatitude SH 500-hPa cyclone and anticyclone. Use the following values and show your work.
   - Planetary vorticity $= -10.3 \times 10^{-5} \, \text{s}^{-1}$.
   - Relative vorticity of the 500-hPa cyclone $= -18 \times 10^{-5} \, \text{s}^{-1}$.
   - Relative vorticity of the 500-hPa anticyclone $= 5 \times 10^{-5} \, \text{s}^{-1}$.
6. You observe the following over Halifax, Nova Scotia. What will be the resulting vertical motion? Explain your answer.
   - Right exit quadrant of a straight jet streak.
   - Strong 500-hPa CVA.
7. You observe the following over Portland, Oregon. What will be the resulting vertical motion and expected weather? Explain your answer.
   - No jet streak.
   - Strong 500-hPa AVA.

8

LOWER-TROPOSPHERIC PROCESSES

In this chapter, we will focus on the lower-tropospheric portion of the three-dimensional circulation in Fig. 7.1 and its applications to weather forecasting. By the conclusion of this chapter, the reader should know how to diagnose regions of lower-tropospheric temperature advection using both 850-hPa and mean sea-level pressure (MSLP) charts. In addition, he or she should be able to evaluate the resultant vertical motion and sensible weather at any point by collectively assessing the three synoptic-scale mechanisms associated with vertical motion: jet streak divergence/convergence, 500-hPa vorticity advection, and lower-tropospheric temperature advection.

8.1 Thickness and Thermal Wind

8.1.1 Thickness: Definition

Recall that the density of the troposphere is proportional to its mean temperature: a cold troposphere is more dense and therefore "shorter" in height from surface to tropopause, while a warm troposphere is less dense and therefore "taller." Because meteorologists primarily use isobaric charts above the surface, it is useful to have a quantity that measures the distance between isobaric surfaces, called *thickness*. Different thickness layers can be useful for various applications, including 1000–700, 1000–850, and 700–400 hPa. Here, however, we will focus on 1000–500-hPa layer thickness and its applications.

Fig. 8.1 shows schematics of two air columns, one cold and one warm. Notice that the 1000–500-hPa layer, indicated by the lowest two blocks and black arrows, is considerably more *thick* in the warm air column than in the cold column. 1000–500-hPa thickness (Δz) is calculated by subtracting the geopotential height at 1000 hPa from the geopotential height at 500 hPa. As a result, thickness has the same units as geopotential height, typically

Synoptic Analysis and Forecasting. https://doi.org/10.1016/B978-0-12-809247-7.00008-9

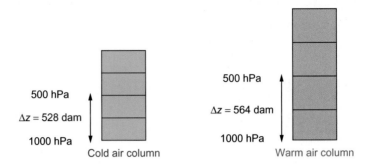

Fig. 8.1 Schematic illustrating the concept of thickness (Δz) in the 1000–500-hPa layer. The cold air column (left) has a smaller thickness (528 dam) than the warm air column (right) (564 dam), because cold air is more dense.

reported in decameters (dam). In the simplistic schematics of Fig. 8.1, the cold air column has a 1000–500-hPa thickness of 528 decameters, while the warm air column's thickness is 564 dam. These thickness values are proportional to the 1000–500-hPa *mean temperature* in each air column. It is important to understand that layer thickness is only proportional to the mean temperature in that layer and not necessarily related to the temperature of a single specific level within that layer (e.g., the surface).

Thickness has several applications, not least of which is the ability to plot it on a weather chart and use as a proxy for the mean temperature of a layer. As we will see later in this chapter, thickness can be extremely useful in helping to diagnose strong temperature gradients, frozen precipitation, and regions of temperature advection.

8.1.2 Thermal Wind Balance

Our first application of thickness is to explore the relationship between lower-tropospheric temperature gradients and the polar jet stream, which is found at the top of the troposphere in the midlatitudes. This relationship is called the *thermal wind relationship*,[1] in which a thickness gradient is related to vertical wind shear. Although we will not quantitatively analyze the thermal wind equation (one of the most important equations in meteorology), we will qualitatively apply its principles in this section.

---

[1] American Meteorological Society, 2012: Glossary of Meteorology, available online at: http://glossary.ametsoc.org/wiki/Thermal_wind.

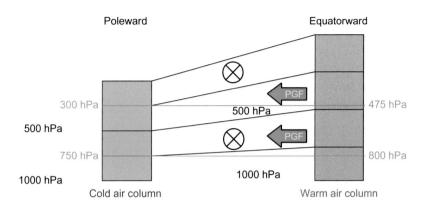

Fig. 8.2 NH schematic of how the thermal wind relationship relates horizontal temperature gradients with the polar jet stream. The cold air column (left) has a smaller thickness than the warm air column (right), because cold air is more dense, with the *green lines* indicating constant altitudes at different pressure levels in each column (*labeled in green*), *green and black arrows* representative of the wind blowing from high to low pressure due to the pressure gradient force (PGF), and *circled black "X"* marks representing the turn of the wind into the page due to the Coriolis Force.

Fig. 8.2 displays the same two air columns as in Fig. 8.1; the poleward column is colder and less thick, while the equatorward column is warmer and thicker. Focus on the two green lines in Fig. 8.2, each indicative of a single altitude; note that each green line is located at a higher pressure in the warm column than in the cold column, because the warm column is less dense. Recall that the pressure gradient force (PGF) initially causes the wind to blow from high to low pressure. The green arrows in Fig. 8.2 illustrate how the PGF initiates air to move from the warm column to the cold column, because of the higher pressures in the warm column at a specific altitude. Next, recall that the Coriolis Force cannot change wind speed, only wind direction. In the Northern Hemisphere (NH), the Coriolis Force turns the wind 90° to the right, which results in the geostrophic wind parallel to geopotential height contours. In the Southern Hemisphere (SH), the Coriolis Force turns the wind 90° to the left, which also results in the geostrophic wind parallel to geopotential height contours. The impact of the Coriolis Force in the NH is highlighted by the circled black "X" marks in Fig. 8.2, indicating that the wind, originally blowing from high to low because of the PGF, makes a right turn "into the page." Flow into the page is representative of a westerly geostrophic wind in the NH.

The diagram in Fig. 8.2 shows how temperature (thickness) gradients are related to the polar jet stream through the thermal wind relationship. Because there is a poleward-equatorward temperature (thickness) gradient, a pressure gradient forms across the

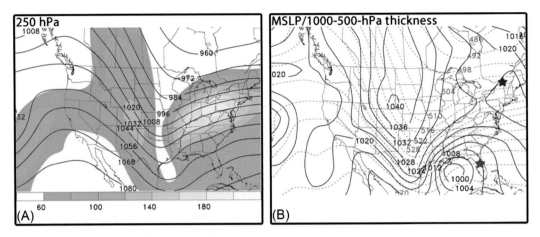

Fig. 8.3 At 00Z 13 March 1993: (A) 250-hPa geopotential height (dam, *solid black contours*) and isotachs (kt, *shaded*), (B) MSLP (hPa, *solid black contours*) and 1000–500-hPa thickness (dam, *dashed red contours*). The *blue and red stars* are discussed in Section 8.1.3 with respect to critical thickness values.

same region. As a result, the wind starts to blow due to the PGF, turns 90° to the right (left) in the NH (SH) due to the Coriolis Force, and becomes the westerly polar jet stream in the upper troposphere. To summarize, the polar jet stream forms due to horizontal temperature (thickness) gradients, and the jet stream's intensity is proportional to the intensity of the temperature (thickness) gradient.

Fig. 8.3 shows 250-hPa and MSLP/1000–500-hPa thickness charts during the development of the 1993 Superstorm. In Chapters 5 and 7, we analyzed both jet streaks seen over the United States in Fig. 8.3A. Here, we want to illustrate the thermal wind relationship by relating each jet streak to a thickness gradient. Focus on the dashed red contours in Fig. 8.3B; these are lines of constant 1000–500-hPa thickness. In the eastern United States, the thickness is ∼510 dam across the eastern Great Lakes and 564 dam in Florida and Gulf of Mexico. This is a 54 dam north-south thickness difference, representative of a strong thickness gradient, which, through the thermal wind relationship, promotes a strong polar jet stream aloft, as evidenced by the eastern US jet streak in Fig. 8.3A.

Across the jet streak over the Rockies (Fig. 8.3A), the thickness is ∼504 dam over the Dakotas, and 552–558 dam over New Mexico and Arizona (Fig. 8.3B). This results in a large thickness gradient, but one not quite as strong as in the eastern United States, resulting in a jet streak that is not quite as strong. The primary take-home point here is that regions of strong thickness gradients are associated with strong winds aloft (i.e., a strong jet stream), while regions of weak thickness gradients (i.e., across Baja California in Fig. 8.3B) are associated with relatively weak winds aloft (Fig. 8.3A).

8.1.3 Critical Thickness Values

The second practical application of 1000–500-hPa thickness is forecasting regions of frozen or mixed phase precipitation (i.e., ice pellets, freezing rain) and the rain-snow line. Because of its relationship to mean column temperature, certain values of thickness have been empirically determined to be representative of the approximate rain-snow line; at a given location, this value is called *critical thickness*. It is important to remember that critical thickness should only be used as an approximation to the rain-snow line and is quite dependent on location, elevation, and season. There are also more modern tools, such as radar (Chapter 12) and soundings (Chapter 14), that can be employed to analyze and forecast frozen precipitation. Nevertheless, to the first order, critical thickness can be useful as a precipitation-type discriminator.

Table 8.1 shows critical thickness values for three different regions across North America. East of the Rockies, 540 dam is the approximate critical thickness value. Regions with 1000–500-hPa thickness <540 dam will generally see snow if precipitation is falling, while areas with 1000–500-hPa thickness > 540 dam will see rain. Mixed precipitation, such as ice pellets and freezing rain, is most likely when the thickness is ~540 dam. There are exceptions to this rule, particularly near large, relatively warm bodies of water such as the Atlantic Ocean and Gulf of Mexico, where slightly smaller thicknesses are typically required to get snow.

Table 8.1 Typical 1000–500-hPa Critical Thickness Values for Various Regions Across North America

Location	Critical 1000–500-hPa Thickness (dam)
East of Rockies	540
Rockies	550
West coast	525

In the Rockies, the critical thickness value (~550 dam) is quite a bit larger than in eastern North America (Table 8.1). Recall that the surface elevation in the Rockies is, on average, located at 800–850 hPa. As a result, a good portion of the 1000–500-hPa layer is below the ground, and surface air at higher elevations is frequently colder than air at sea level. This means that in regions of high terrain, larger 1000–500-hPa thicknesses can still support snow. In fact, many autumn and spring snowstorms in places such as Denver and Calgary occur with 1000–500-hPa thickness values of 545–550 dam.

On North America's west coast, the Pacific Ocean is a major source of air mass moderation. Although the water temperature is not particularly warm (5–15°C), it is usually still warmer than the adjacent land mass during the winter. Therefore the boundary layer is typically relatively warm and thick, meaning that the 1000–500-hPa thickness must be relatively small in order for snow to fall. It is rare for snow to occur near the Pacific Ocean (e.g., Vancouver, Seattle, Portland) with thicknesses >525 dam. It is crucial that the analyst or forecaster know regional differences and not blindly apply the well-known "540-dam line" as the critical thickness value!

Fig. 8.3 shows a blue star near Montreal, Quebec and a red star in central Florida. Interpolating between lines of constant thickness, the 1000–500-hPa thickness at Montreal is ~507 dam, while in central Florida it is roughly 562 dam. Because both locations are east of the Rockies, we can use 540 dam as the approximate critical thickness value. Therefore we can conclude that if precipitation were to fall at Montreal, it would fall as snow, while precipitation in central Florida would fall as rain. Remember that critical thickness alone cannot determine whether precipitation will fall; it is merely a clue to what *type* of precipitation may occur.

8.2 Temperature Advection

8.2.1 Definition

In Chapter 7, we discussed two upper-tropospheric mechanisms associated with vertical motion: jet streak divergence/convergence and 500-hPa geostrophic vorticity advection. Here, we introduce the third synoptic-scale mechanism for vertical motion: *lower-tropospheric geostrophic temperature advection*. Along the same lines as geostrophic vorticity advection (Chapter 7), geostrophic temperature advection is the transport of temperature by the geostrophic wind. Temperature advection tends to be quite small in the upper troposphere, where the isohypses and isotherms are mostly parallel to each other, so we will focus on the lower troposphere.

In Chapter 6, we paid special attention to regions on 850-hPa charts where the isohypses and isotherms were nearly perpendicular to each other; this is a key clue to recognize regions of temperature advection. Recall that the geostrophic wind blows parallel to isohypses (on an isobaric chart) and parallel to MSLP isobars on a surface chart. To identify temperature advection, the analyst or forecaster should ask "what type of air is the geostrophic wind transporting into a given region: warm or cold?" Fig. 8.4A shows a schematic of the geostrophic wind parallel to the isotherms (thickness lines); in this case, there is no temperature advection. However, in Fig. 8.4B, the geostrophic wind is perpendicular to the isotherms (thickness lines) and blowing from warm to cold; this is defined as *warm-air advection* (WAA). In contrast, Fig. 8.4C shows the geostrophic wind perpendicular to the isotherms (thickness lines), but blowing from cold to warm; this is defined as *cold-air advection* (CAA).

Now that we understand how to recognize WAA and CAA, let us explore how each relates to vertical motion. During WAA, the

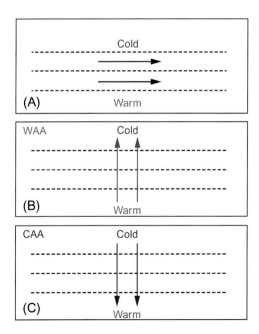

Fig. 8.4 Schematic illustrating the concept of lower-tropospheric temperature advection for (A) no temperature advection, (B) warm-air advection (WAA), and (C) cold-air advection (CAA). The *black dashed lines* represent isotherms (or *thickness lines*), and the *arrows* represent the geostrophic wind.

Table 8.2 Lower-Tropospheric Temperature Advection and Its Association With Divergence/Convergence and Vertical Motion

	WAA	CAA
Location	Ahead of a cyclone/behind an anticyclone	Ahead of an anticyclone/behind a cyclone
Surface air motion	Convergence	Divergence
Vertical motion	Ascent	Descent

lower troposphere warms as warm air is advected into a region from somewhere else. This disrupts thermal wind balance, which the atmosphere immediately tries to restore. To do this, the atmosphere must cool adiabatically, which is accomplished through ascent. Through the mass continuity principle, ascent must be accompanied by surface convergence (Chapter 7). In summary, lower-tropospheric WAA is associated with ascent and surface convergence (Table 8.2).

During CAA, the lower troposphere cools as cold air is advected in from elsewhere. This also disturbs thermal wind balance, which the atmosphere immediately tries to restore through adiabatic warming by descent. Through the mass continuity principle, descent will be accompanied by surface divergence (Chapter 7). In summary, lower-tropospheric CAA is associated with descent and surface divergence (Table 8.2).

8.2.2 Identification

In regions of relatively low elevation (surface pressure > 850 hPa), 850-hPa and MSLP/1000–500-hPa thickness charts can largely be used interchangeably to identify regions of geostrophic temperature advection. On an 850-hPa chart, the geostrophic wind is always parallel to the isohypses, while on a MSLP/1000–500-hPa thickness chart, the geostrophic wind is always parallel to the isobars. Fig. 8.5 shows 850-hPa (Fig. 8.5A) and MSLP/1000–500-hPa thickness (Fig. 8.5B) charts from 12Z 13 March 1993, during the height of the Superstorm. A strong surface cyclone is centered over the Georgia/South Carolina border. Note that the isotherms (thickness lines) in both panels are

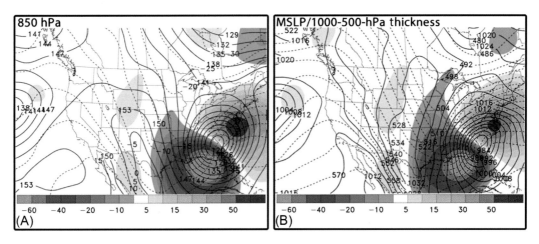

Fig. 8.5 At 12Z 13 March 1993: (A) 850-hPa geopotential height (dam, *solid black contours*), temperature (°C, *dashed black contours*), and geostrophic horizontal temperature advection ($\times 10^{-5}$ K s^{-1}, *shaded warm colors* for WAA, *cool colors* for CAA), (B) MSLP (hPa, *solid black contours*), 1000–500-hPa thickness (dam, *dashed black contours*), and 1000–700-hPa geostrophic horizontal temperature advection ($\times 10^{-5}$ K s^{-1}, *shaded warm colors* for WAA, *cool colors* for CAA).

dashed black contours, so as not to be obscured by the red shading used for WAA.

Focus on the region to the north and northeast of the surface cyclone center across Delaware, Maryland, Virginia, and the Carolinas (Fig. 8.5). On the 850-hPa chart (Fig. 8.5A), both the isohypses and isotherms are tightly bunched; the same is true of the isobars and thickness lines on the MSLP chart (Fig. 8.5B). These assessments indicate a strong lower-tropospheric temperature gradient. The geostrophic wind is east and southeasterly, and largely perpendicular to the isotherms (thickness lines), meaning that the wind is blowing from warm to cold across a strong temperature (thickness) gradient. Consequently, this is a region of strong WAA, as indicated by the orange and red shading. The strongest WAA in each panel is located where there is a combination of tightly spaced and perpendicular isohypses/isotherms (Fig. 8.5A) or isobars/thickness lines (Fig. 8.5B). Strong WAA implies surface convergence and forcing for ascent, which leads to clouds and precipitation (Table 8.2). WAA is most commonly located ahead (downstream) of a surface cyclone and behind (upstream) of a surface anticyclone (Table 8.2).

Now focus on the Gulf Coast region in both panels in Fig. 8.5. Here, behind the surface cyclone and ahead of the strong surface anticyclone located over the central United States (Chapter 6), the geostrophic wind is predominantly north and northwesterly. Both the isohypses (isobars) and isotherms (thickness lines) are tightly spaced and perpendicular to each other, so air blows from cold to warm. Therefore the blue- and purple-shaded region indicates CAA, surface divergence, and forcing for descent (Table 8.2). Using temperature advection alone, skies should be generally clear in a strong CAA region, as temperatures cool behind the cyclone. CAA is most commonly located ahead (downstream) of a surface anticyclone and behind (upstream of) a surface cyclone (Table 8.2). There are a few other regions of weak temperature advection evident in Fig. 8.5, such as the WAA over the central Rockies and the CAA in northeastern Quebec. As we previously discussed, lower-tropospheric temperature advection is only one of three synoptic-scale forcing mechanisms for vertical motion, and thus presents an incomplete picture when used alone. However, it is important to consider temperature advection with the other two mechanisms (jet streak divergence/convergence and 500-hPa vorticity advection), in order to derive a complete assessment of vertical motion at any point. Analyzing the sign and intensity of all three mechanisms at the same time is a fantastic way to diagnose the current or future weather conditions at a given location. We discuss this three-dimensional approach to predict the intensification and motion of weather systems, in Chapter 9.

8.3 Questions and Practice Exercises

1. Using the thermal wind relationship, explain why the polar jet stream in the SH is primarily *westerly* (as in the NH). Draw a diagram if it helps.
2. For the following locations and reported thicknesses on a winter day, use your knowledge of 1000–500-hPa critical thicknesses to determine the most likely type of precipitation. Explain your answers.
   - Philadelphia, Pennsylvania: 541 dam.
   - Kelowna, British Columbia: 549 dam.
   - Eureka, California: 533 dam.
3. Using the MSLP/1000–500-hPa thickness chart in Fig. 8.3B, identify the type of temperature advection at each of the following locations:
   - Wichita, Kansas.
   - Savannah, Georgia.
   - Ottawa, Ontario.

4. Using the MSLP/1000–500-hPa thickness chart in Fig. 8.5B, which of the following regions *should* feature a stronger jet streak at the tropopause? Explain your answer.
   - Oregon, Washington, British Columbia.
   - The northeastern United States and southeastern Quebec.
5. It is a February day in Frankfurt, Germany. Using your usual assortment of weather charts, you observe the following mechanisms. Describe the expected sense of vertical motion and resultant weather.
   - Left entrance region of a 250-hPa jet streak.
   - No 500-hPa vorticity advection.
   - Strong lower-tropospheric CAA.
6. It is an October day in Bermuda. Using your usual assortment of weather charts, you observe the following mechanisms. Describe the expected sense of vertical motion and resultant weather.
   - Right entrance region of a 250-hPa jet streak.
   - Strong 500-hPa CVA.
   - Weak lower-tropospheric WAA.

9

PUTTING IT ALL TOGETHER

9.1 Mechanisms Associated With Vertical Motion

We have previously identified three forcing mechanisms associated with vertical motion: jet streak divergence, upper-tropospheric (e.g., 500-hPa) vorticity advection, and lower-tropospheric (e.g., 850-hPa) temperature advection. Table 9.1 summarizes the processes that are associated with ascent (i.e., clouds and precipitation) and descent (i.e., clear skies and fair weather). It is important to remember that the atmosphere does not always produce textbook examples; mechanisms can contradict each other, leaving the sign of vertical motion dependent on the relative magnitude of each forcing mechanism.

Table 9.2 shows common combinations of forcing mechanisms, the resulting vertical motion, and likely expected weather. For the unambiguous ascent case [i.e., divergent region of a jet streak, cyclonic vorticity advection (CVA), and warm-air advection (WAA)], one can expect clouds and precipitation. In a different scenario, two of the three mechanisms (e.g., CVA, WAA) may indicate ascent, while the third mechanism is neutral or associated with descent (e.g., convergent region of a jet streak). This setup indicates weak-to-moderate ascent, suggesting mostly cloudy skies and a chance of precipitation. Because the magnitude of the ascent is ambiguous, the forecaster should try to quantitatively assess the magnitude of each mechanism.

Fig. 9.1 details the evolution of the Groundhog Day Blizzard of 2011, which caused up to 2 ft. of snow in the central United States, and severe weather along the Gulf Coast, in the warm sector of the cyclone. Here, we will focus on Chicago, Illinois and Fort Smith, Arkansas.

Fig. 9.1A shows that at 00Z, Chicago is located in a dual jet streak divergent region: the exit region of the curved jet to the southwest and the right entrance region of the straight jet to

Synoptic Analysis and Forecasting. https://doi.org/10.1016/B978-0-12-809247-7.00009-0

Table 9.1 Summary of the Synoptic-Scale Mechanisms Associated With Vertical Motion

	Ascent	Descent
Jet streak divergence	Divergent region	Convergent region
Upper-tropospheric vorticity advection	Cyclonic vorticity advection (CVA)	Anticyclonic vorticity advection (AVA)
Lower-tropospheric temperature advection	Warm-air advection (WAA)	Cold-air advection (CAA)

Table 9.2 Examples of Common Mechanism Combinations for Vertical Motion

Jet Streak Divergence	Vorticity Advection	Temperature Advection	Vertical Motion	Likely Weather
Divergent region	CVA	WAA	Strong ascent	Clouds and precipitation
Convergent region	CVA	WAA	Ascent	Mostly cloudy, chance of precipitation
Neutral	CVA	CAA	Depends on relative magnitudes of CVA and CAA	Depends on relative magnitudes of CVA and CAA
Divergent region	AVA	CAA	Descent	Partly cloudy, no precipitation
Convergent region	AVA	CAA	Strong descent	Clear, no precipitation
Neutral	AVA	WAA	Depends on relative magnitudes of AVA and WAA	Depends on relative magnitudes of AVA and WAA

Listed are the three synoptic-scale mechanisms associated with vertical motion, the resulting sign of vertical motion, and the most likely expected weather.

the northeast. At 500-hPa (Fig. 9.1B), Chicago is located in a region of CVA downstream of the strong 500-hPa trough centered over southern Missouri. Finally, at 850-hPa (Fig. 9.1C), Chicago is located in a region of strong WAA ahead of the surface cyclone centered over western Illinois. Therefore all three forcing

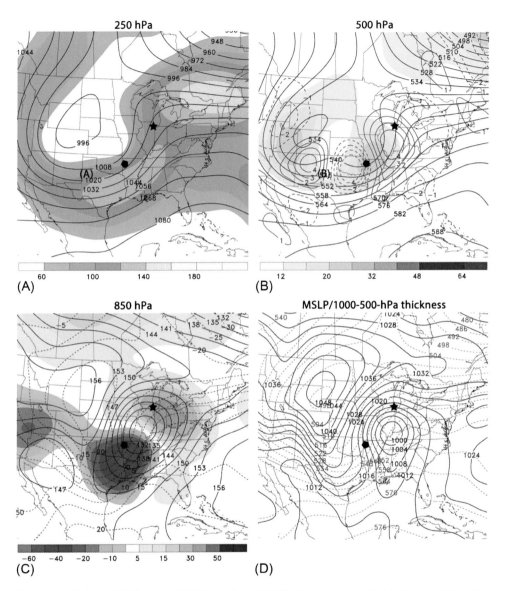

Fig. 9.1 Case example from 00Z 2 February 2011. Plotted are (A) 250-hPa geopotential height (dam, *solid black contours*) and wind speed (kt, *shaded*), (B) 500-hPa geopotential height (dam, *solid black contours*), geostrophic absolute vorticity ($\times 10^{-5}$ s^{-1}, *shaded*), and geostrophic absolute vorticity advection ($\times 10^{-9}$ s^{-2}, *blue contours, solid* for CVA, *dashed* for AVA), (C) 850-hPa geopotential height (dam, *solid black contours*), temperature (°C, *dashed black contours*), and geostrophic horizontal temperature advection ($\times 10^{-5}$ K s^{-1}, *shaded warm colors* for WAA, *cool colors* for CAA), and (D) MSLP (hPa, *solid black contours*) and 1000–500-hPa thickness (dam, *dashed red contours*). Chicago, Illinois and Fort Smith, Arkansas are marked in each panel with a *black star* and *pentagon*, respectively.

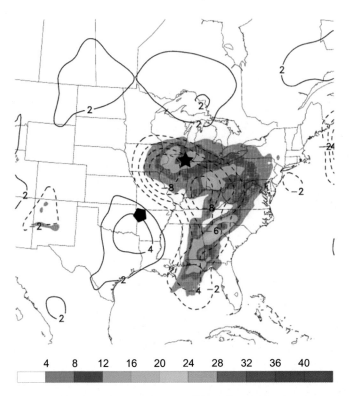

Fig. 9.2 850–500-hPa layer-averaged vertical motion (ω, $\times 10^{-3}$ Pa s^{-1}, *solid black contours* for descent, *dashed black contours* for ascent) at 00Z 2 February 2011, and 6-h accumulated precipitation (mm, *shaded*) between 00 and 06Z 2 February 2011. Chicago, Illinois and Fort Smith, Arkansas are marked with a *black star* and *pentagon*, respectively.

mechanisms indicate ascent at Chicago. Fig. 9.2 confirms this by showing large negative values of vertical motion (i.e., ascent) and 16–24 mm of accumulated liquid-equivalent precipitation between 00 and 06Z over Chicago.

Fig. 9.1A shows Fort Smith located near the center, but still in the entrance region of a cyclonically curved jet streak, indicative of weak upper-tropospheric convergence. At 500-hPa, Fort Smith is in an area of strong anticyclonic vorticity advection (AVA, Fig. 9.1B), while at 850-hPa an area of strong cold-air advection (CAA) is present (Fig. 9.1C). Therefore all three forcing mechanisms indicate descent over Fort Smith, which is confirmed by positive values of vertical motion (i.e., descent) and no precipitation (Fig. 9.2). The forecaster can thus be confident in a prediction of mostly clear skies and decreasing temperatures due to CAA.

9.2 Motion and Intensity of Upper-Tropospheric Troughs and Ridges

The motion and intensification of upper-tropospheric (i.e., 500-hPa) troughs and ridges can largely be explained by upper-tropospheric vorticity advection and lower tropospheric temperature advection, respectively. Although the associated equations are complex, these processes can be understood qualitatively, allowing the forecaster to predict how 500-hPa troughs and ridges will evolve over time. While we will briefly discuss the impacts of jet streaks on the intensification of 500-hPa troughs, they generally play a minor role compared to the influence of vorticity and temperature advections.

9.2.1 Motion

Recall that a 500-hPa trough is a geostrophic absolute vorticity maximum, while a 500 hPa ridge is a geostrophic absolute vorticity minimum. To assess geostrophic vorticity advection (regions of CVA and AVA), the forecaster must ask "to where is the geostrophic wind transporting the maximum or minimum vorticity?" Because troughs are cyclonic relative vorticity maxima, they preferentially move toward regions of CVA. Likewise, ridges are anticyclonic relative vorticity maxima, so they preferentially move toward regions of AVA. These assertions assume that relative vorticity advection is of a larger magnitude than planetary vorticity advection, which is generally the case for all but the longest wavelength troughs and ridges. For the rest of this chapter, we will assume that the magnitude of relative vorticity advection exceeds that of planetary vorticity advection.

Fig. 9.3 shows the evolution of the 500-hPa flow during the 2011 Groundhog Day. Blizzard. In Fig. 9.3A, two trough axes (dashed blue lines) and one ridge axis (jagged red line) are highlighted. We would expect the troughs to move toward regions of CVA and the ridge toward AVA. For the southern trough located over New Mexico and Arizona at 00Z 1 February (Fig. 9.3A), the strongest CVA is located to the east, which is where the trough proceeds to move over the next 12–24 h (Fig. 9.3B and C). The northern trough initially located over Montana (Fig. 9.3A) also moves toward CVA, which is located to its south (Fig. 9.3A and B). Meanwhile, the ridge axis located over Alberta and Saskatchewan at 00Z 1 February moves toward the AVA located to its south-southwest (Fig. 9.3A and B). As the flow pattern intensifies over the central United States, the 500-hPa trough formerly located over

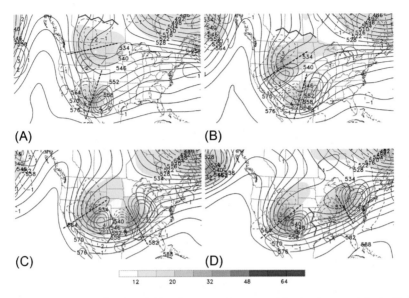

(A) (B)

(C) (D)

| 12 | 20 | 32 | 48 | 64 |

Fig. 9.3 500-hPa geopotential height (dam, *solid black contours*), geostrophic absolute vorticity ($\times 10^{-5}$ s^{-1}, *shaded*), and geostrophic absolute vorticity advection ($\times 10^{-9}$ s^{-2}, *blue contours*, solid for CVA, *dashed* for AVA) for (A) 00Z 1 February 2011, (B) 12Z 1 February 2011, (C) 00Z 2 February 2011, and (D) 12Z 2 February 2011. 500-hPa trough (ridge) axes are marked with *dashed blue* (*jagged red*) *lines* in each panel.

the southwestern United States moves northeastward in the direction of the largest CVA (Fig. 9.3B–D). Accordingly, the shortwave ridge located over Oklahoma and Texas at 00Z 2 February moves east-northeastward in the direction of AVA (Fig. 9.3C and D).

Using the rule of thumb that troughs move toward CVA and ridges move toward AVA can help the forecaster predict where each feature will move over time. This can be done qualitatively without plotting vorticity advection, by evaluating where the geostrophic wind (which is always parallel to the geopotential height contours) will transport vorticity maxima and minima. In the midlatitudes, because the upper-tropospheric flow is generally westerly, CVA (AVA) is often but not always located to the east of a 500 hPa trough (ridge).

Recall that 500-hPa closed and cutoff cyclones typically move eastward very slowly and can even slightly retrogress (move westward). Fig. 9.4 shows the 3-day evolution of two 500-hPa cutoff cyclones in May 2013, one located over California and the other over the southeastern United States. The polar jet stream during this time period is located in central Canada, with various shortwave troughs and ridges embedded in the westerly flow. On the other hand, the two cutoff cyclones over the southern United States drift very slowly downstream over the 3-day period.

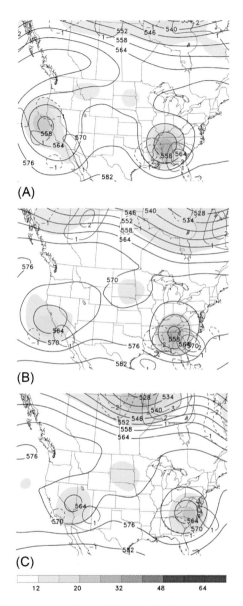

Fig. 9.4 500-hPa geopotential height (dam, *solid black contours*), geostrophic absolute vorticity ($\times 10^{-5}$ s^{-1}, *shaded*), and geostrophic absolute vorticity advection ($\times 10^{-9}$ s^{-2}, *blue contours, solid* for CVA, *dashed* for AVA) for (A) 00Z 6 May 2013, (B) 00Z 7 May 2013, (C) 00Z 8 May 2013.

We have already seen that 500-hPa troughs move toward regions of CVA. In the case of 500-hPa closed and cutoff cyclones, CVA is typically (a) weaker and (b) rotating around the cyclone itself. This can be explained by recalling that the geostrophic wind is always parallel to the geopotential height contours; thus cyclonic relative vorticity is advected in a circular fashion around the cutoff cyclone. Therefore, the cyclone drifts slowly because the CVA is colocated with the cyclone. This type of evolution is shown in Fig. 9.4; over the course of 48 h, the western cutoff cyclone slowly moves from the central California coast to southern Nevada, while the eastern cutoff cyclone slowly advances from Alabama/Tennessee to North Carolina. Qualitatively comparing the rate of motion of each cutoff cyclone with that of the embedded troughs and ridges in the polar jet stream in central Canada, it is clear that the cutoff cyclones move much more slowly.

Fig. 9.4 also shows that the CVA associated with both cutoff cyclones rotates cyclonically around each cyclone as time moves forward. As an example, for the eastern cyclone the CVA is initially located to the east-southeast of the vorticity maximum (Fig. 9.4A), but the CVA rotates counterclockwise to the east-northeast of the cyclone over the course of 48 h (Fig. 9.4C). For practical forecast applications, these systems and the ascent associated with CVA can impact a particular region for up to a week. The more slowly a cyclone moves, the greater potential it has to cause hazardous weather for a longer period of time. Numerous infamous flood events, such as the 2013 Alberta[1] and Great Colorado[2] floods, were associated with cutoff cyclones.

9.2.2   Intensification and Decay

9.2.2.1   Role of Temperature Advection

Recall from Chapter 8 that thickness is the geopotential height difference between two pressure levels, in practice most commonly 1000 and 500 hPa. Larger (smaller) 1000–500-hPa thickness indicates:

- A larger (smaller) distance between the 1000- and 500-hPa height surfaces.
- Warmer (colder) average temperature in the 1000–500-hPa layer.

[1] Milrad, S. M., J. R. Gyakum, and E. H. Atallah, 2015: A meteorological analysis of the 2013 Alberta Flood: Antecedent large-scale flow pattern and synoptic-dynamic characteristics. *Mon. Wea. Rev.,* **143,** 2817–2841.
[2] Gochis, D., and Coauthors, 2015: The Great Colorado Flood of September 2013. *Bull. Amer. Meteor. Soc.,* **96,** 1461–1487.

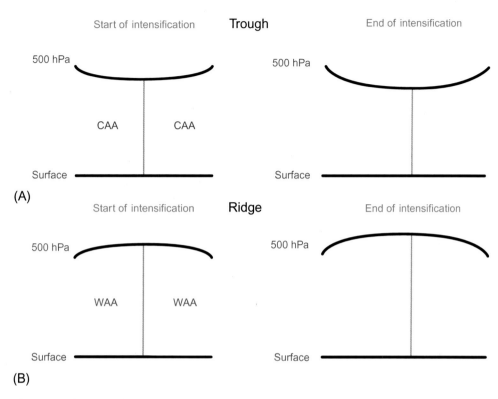

Fig. 9.5 Schematic demonstrating lower-tropospheric temperature advection affects the intensity of upper-tropospheric (500-hPa) (A) troughs and (B) ridges, using the concept of thickness (*orange lines*).

We will now use the concept of thickness to help explain the intensification of 500-hPa troughs and ridges.

Fig. 9.5 shows schematics of a 500-hPa trough and 500-hPa ridge that both intensify over time. The "surface" is assumed to be in the lower troposphere (e.g., 1000-hPa), dependent on location elevation. In the trough case (Fig. 9.5A), the 500-hPa height surface has an initial cyclonic curvature, with the orange line representing the initial surface 500-hPa thickness. If there is lower-tropospheric (e.g., 850-hPa) CAA underneath the 500-hPa trough, it cools the surface-500-hPa layer and reduces the thickness. Because the surface is a boundary condition, most of the reduction in column thickness is seen at the 500-hPa level; that is, for thickness to decrease, 500-hPa heights must fall, resulting in a stronger 500-hPa trough. An equal but opposite process results in 500-hPa ridge intensification. Fig. 9.5B shows an initial 500-hPa ridge with lower-tropospheric WAA underneath it. WAA acts to warm the column and increase thickness, causing 500-hPa heights to rise and therefore intensifying the 500-hPa ridge.

The following conclusions arise from these thickness-based arguments:

- Lower-tropospheric CAA → column thickness decreases → 500-hPa heights fall → stronger (weaker) 500-hPa trough (ridge).
- Lower-tropospheric WAA → column thickness increases → 500-hPa heights rise → stronger (weaker) 500-hPa ridge (trough).

Fig. 9.6 shows the evolution of the February 2011 Groundhog Day Blizzard during the period of most rapid intensification. At 12Z 1 February (Fig. 9.6A), a 500-hPa trough is located over Texas, while a ridge axis is located downstream over Iowa and eastern Missouri. At 850 hPa (Fig. 9.6B), a cyclone is located over eastern Oklahoma, with a region of WAA to its northeast and CAA to its southwest. The WAA is located directly underneath the 500-hPa ridge axis, while the CAA is located underneath and slightly downwind of the 500-hPa trough axis. Therefore, the 500-hPa heights should fall near and just east of the trough axis and rise in the ridge axis. This is confirmed 6 h later (18Z 1 February) in Fig. 9.6B, where the trough over Oklahoma and Texas and ridge over Illinois and Indiana have both become more amplified. At this time, there is still WAA underneath the ridge axis and CAA underneath the trough axis (Fig. 9.6E), suggesting that both features will continue to intensify. At 00Z 2 February (Fig. 9.6C), both the trough and ridge have amplified further. They continue to do so (not shown) until the trough catches up to the surface cyclone, such that the WAA and CAA are no longer located underneath the 500-hPa ridge and trough axis, respectively. Finally, note that the trough's tilt evolves from slightly positive to negative over 12 h (Fig. 9.6A–C). This occurs because of the lower-tropospheric CAA located below, but just downwind (east) of the 500-hPa trough axis, which causes 500-hPa geopotential heights to fall. The evolution of a trough from positively tilted to negatively tilted is also indicative of an intensifying system.

9.2.2.2 Role of Jet Streak Processes

Although lower-tropospheric temperature advection is the largest factor in determining 500-hPa trough and ridge intensity, jet streak processes can also play a role. Fig. 9.7 shows 250 and 500-hPa plots for the March 1993 Superstorm,[3] which produced profound impacts (blizzard conditions, severe weather, etc.) along

[3] Kocin, P., P. M. Schumacher, R. Morales Jr., and L. Uccellini, 1995: Overview of the 12–14 March 1993 Superstorm. *Bull. Amer. Meteor. Soc.*, **76**, 165–182.

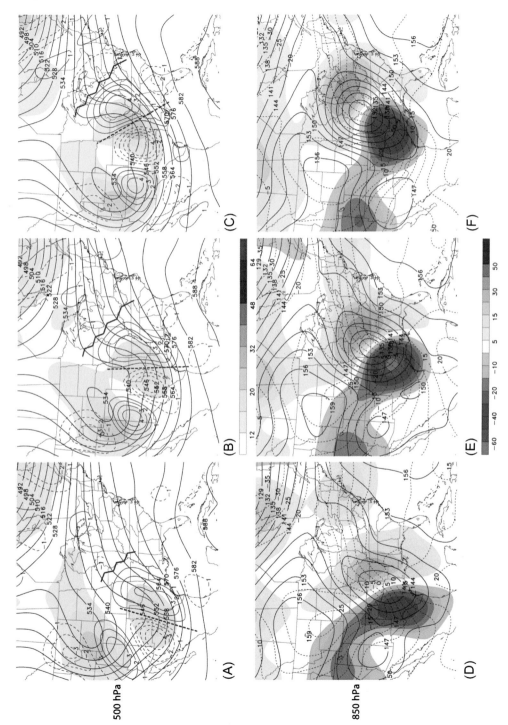

Fig. 9.6 (A–C) 500-hPa geopotential height (dam, *solid black contours*), geostrophic absolute vorticity (×10⁻⁵ s⁻¹, *shaded*), and geostrophic absolute vorticity advection (×10⁻⁹ s⁻², *blue contours, solid* for CVA, *dashed* for AVA), (D–F) 850-hPa geopotential height (dam, *solid black contours*), and geostrophic horizontal temperature advection (×10⁻⁵ K s⁻¹, *shaded warm colors* for WAA, *cool colors* for CAA), for (left) 12Z 1 February 2011, (middle) 18Z 1 February 2011, and (right) 00Z 2 February 2011. In (A–C), 500-hPa trough (ridge) axes are marked with *dashed blue (jagged red) lines* in each panel.

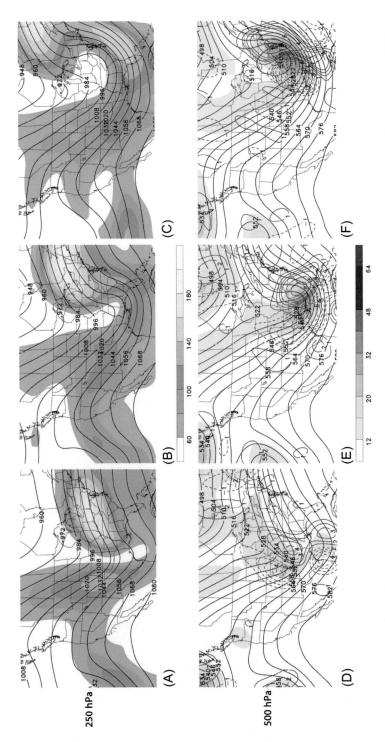

Fig. 9.7 (A–C) 250-hPa geopotential height (dam, *solid black contours*) and wind speed (kt, *shaded*), (D–F) 500-hPa geopotential height (dam, *solid black contours*), geostrophic absolute vorticity ($\times10^{-5}$ s^{-1}, shaded), and geostrophic absolute vorticity advection ($\times10^{-9}$ s^{-2}, *blue contours, solid* for CVA, *dashed* for AVA), for (left) 00Z 13 March 1993, (middle) 12Z 13 March 1993, and (right) 00Z 14 March 1993.

the entire eastern seaboard and set many still-existing MSLP records for an extratropical cyclone over the eastern United States.

The incipient 500-hPa trough that eventually helped to produce the Superstorm was located over the western Gulf of Mexico at 00Z 13 March, while a 140-kt jet streak was located upstream over the Rockies (Fig. 9.7A and D). Because the fastest winds in a jet streak are located in the core, the poleward side of the jet streak is a region of strong cyclonic shear vorticity. In addition, jet streaks tend to move more quickly than 500-hPa troughs and ridges because the winds are usually faster at the tropopause than at 500 hPa. In the Superstorm case, the upstream jet streak quickly moves into the base of the trough between 00 and 12Z 13 March (Fig. 9.6A and B). It is not a coincidence that the 500-hPa trough intensifies (height falls and absolute vorticity increases, Fig. 9.6D–F) during this time period. Although trough intensification is also due to lower-tropospheric CAA, the "injection" of shear vorticity into the trough by the upstream jet streak should not be ignored, particularly in cases where the upstream jet streak is strong (>100 kt).

9.3 Motion and Intensity of Surface Cyclones and Anticyclones

The motion and intensification of lower-tropospheric (surface) cyclones and anticyclones can be explained by temperature and vorticity advection, respectively. Jet streaks can also play a role in surface cyclone intensification by providing a source of upper-level divergence.

9.3.1 Motion

Recall that a surface cyclone is a pressure (height) minimum and a surface anticyclone is a pressure (height) maximum. Furthermore, air converges at the center of a surface cyclone and diverges outward from the center of a surface anticyclone. Surface features move to regions of like pressure tendencies and divergence; specifically, surface cyclones move toward regions of surface pressure falls and convergence, while surface anticyclones move toward regions of surface pressure rises and divergence. We can associate these mechanisms with lower-tropospheric horizontal temperature advection by recalling that WAA is associated with ascent while CAA is associated with descent. Using the mass continuity principle, in which ascent indicates surface convergence and descent indicates surface divergence, surface cyclones move toward regions of WAA and surface anticyclones move toward regions of CAA.

To assess temperature advection on either a lower-tropospheric (e.g., 850-hPa) isobaric chart or a MSLP/1000–500-hPa thickness chart, the forecaster must ask "to where is the geostrophic wind transporting warm and cold air?" Upon doing this, the forecaster can identify regions of WAA and CAA, and predict the motion of surface cyclones and anticyclones.

Fig. 9.8 shows 850-hPa and MSLP/1000–500-hPa thickness charts for the 2011 Groundhog Day Blizzard, with a surface cyclone initially located over eastern Oklahoma and an anticyclone over southern Montana. Although the 850-hPa (Fig. 9.8A–C) and MSLP (Fig. 9.8D–F) features are slightly offset from each other in terms of location, they will each track toward the same type of temperature advection. The surface cyclone initially located over the south-central United States will track northeastward toward WAA, while the surface anticyclone initially located in Montana will slide slowly south-southeastward toward CAA (Fig. 9.8).

The speed of motion is proportional to the strength of the horizontal temperature advection. Strong WAA is evident ahead of the cyclone as the isoheights (isobars) are nearly perpendicular to the strong temperature (thickness) gradient. This suggests that the cyclone will move relatively quickly toward the northeast. In Colorado and Wyoming, the isobars and thickness contours are nearly parallel to each other, and there is relatively weak CAA to the south-southeast (Fig. 9.8A and D). In contrast to the cyclone, this suggests the anticyclone will move relatively slowly toward the south-southeast.

9.3.2 Intensification and Decay

While the differential equations that explain the intensification and decay of surface systems are quite complex, the qualitative concepts are relatively simple. First, we return to the mass continuity principle. Fig. 9.9A shows a situation in which the magnitude of upper-tropospheric divergence exceeds that of surface convergence. This indicates that the mass of the air column decreases over time, which causes surface pressure to decrease and the surface cyclone to intensify. The inverse is true for a surface anticyclone: Fig. 9.9B shows upper-tropospheric convergence exceeding surface divergence, which acts to increase surface pressure and intensify the surface anticyclone. In summary, upper-tropospheric divergence (convergence) above a surface cyclone will intensify (weaken) the cyclone, while upper-tropospheric convergence (divergence) above a surface anticyclone will intensify (weaken) the anticyclone.

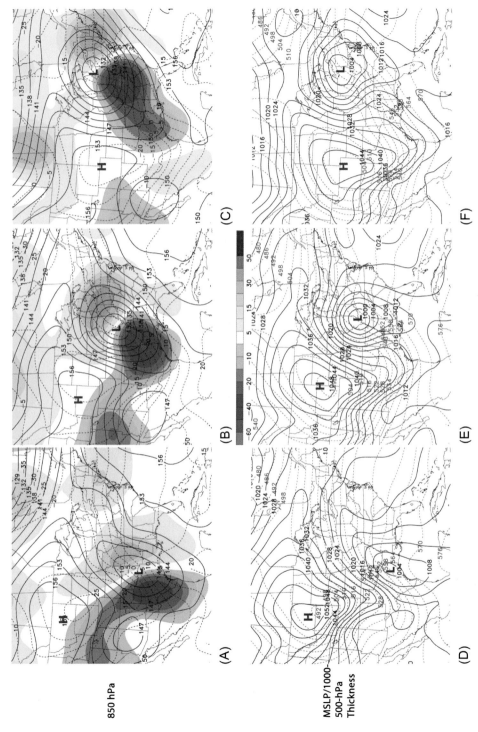

Fig. 9.8 (A–C) 850-hPa geopotential height (dam, *solid black contours*), temperature (°C, *dashed black contours*), and geostrophic horizontal temperature advection (×10⁻⁵ K s⁻¹, *shaded warm colors* for WAA, *cool colors* for CAA), (D–F) MSLP (hPa, *solid black contours*) and 1000–500-hPa thickness (dam, *dashed red contours*), for (left) 12Z 1 February 2011, (middle) 00Z 2 February 2011, and (right) 12Z 2 February 2011. Lower-tropospheric (surface) cyclones and anticyclones are marked with a *red* "L" and *blue* "H," respectively, in each panel.

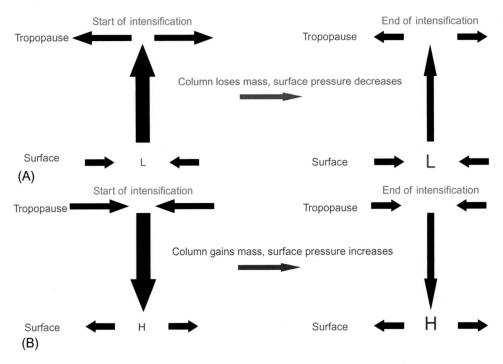

Fig. 9.9 Schematic of how upper-tropospheric divergence and convergence act to intensify lower-tropospheric (surface) (A) cyclones and (B) anticyclones.

Divergent regions of jet streaks and CVA ahead of 500-hPa troughs are both associated with upper-level divergence. Conversely, upper-level convergence is found in convergent regions of jet streaks and colocated with AVA downstream of 500-hPa ridges. Therefore in order to predict the intensification and decay of surface cyclones and anticyclones, the forecaster should examine both jet stream level (e.g., 250-hPa) and 500-hPa charts.

Let us revisit the 1993 Superstorm case of 13–14 March 1993. In the 24 h between 00Z 13 March and 00Z 14 March, the surface cyclone deepened 28 hPa as it moved up the eastern seaboard. In order for a surface cyclone to intensify rapidly, strong upper-level divergence must be present above the cyclone center. At 00Z 13 March 1993, the Superstorm is located underneath the right entrance region of a strong jet streak (Fig. 9.10A) and in a region of CVA ahead of the 500-hPa trough over the Gulf of

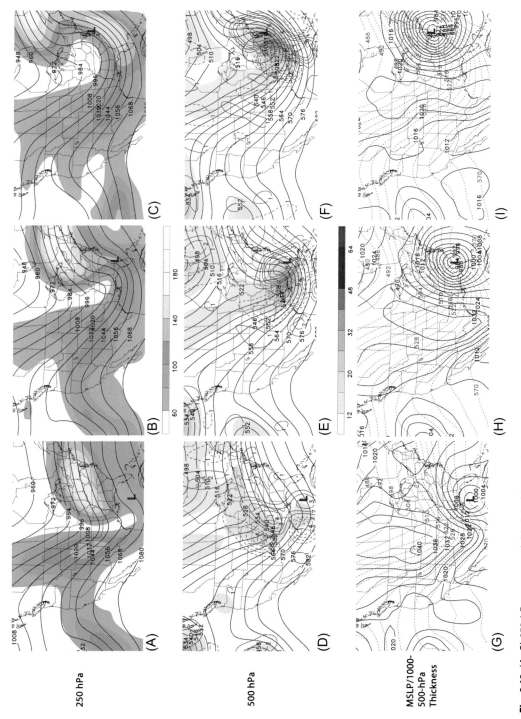

Fig. 9.10 (A–C) 250-hPa geopotential height (dam, *solid black contours*) and wind speed (kt, *shaded*), (D–F) 500-hPa geopotential height (dam, *solid black contours*), geostrophic absolute vorticity ($\times 10^{-5}$ s^{-1}, *shaded*), and geostrophic absolute vorticity advection ($\times 10^{-9}$ s^{-2}, *blue contours, solid* for CVA, *dashed* for AVA), (G–I) MSLP (hPa, *solid black contours*) and 1000–500-hPa thickness (dam, *dashed red contours*), for (left) 00Z 13 March 1993, (middle) 12Z 13 March 1993, and (right) 00Z 14 March 1993. Surface cyclone locations are marked with a *red "L"* in each panel.

Mexico (Fig. 9.10D). Twelve hours later (12Z 13 March), the surface cyclone has intensified by approximately 15 hPa (Fig. 9.10G and H) and is now located in the exit (divergent) region of a cyclonically curved jet streak over the Georgia coast (Fig. 9.10B) and strong CVA ahead of the negatively tilted 500-hPa trough (Fig. 9.10E). Thus, the surface cyclone remains in an area of strong upper-level divergence and will continue to intensify. By 00Z 14 March, the cyclone is located over the Delmarva Peninsula and the sea-level pressure has fallen approximately another 13 hPa. The surface cyclone is still located in the divergent exit region of a cyclonically curved jet streak (Fig. 9.10C), but the 500-hPa trough is now directly above the surface cyclone. Therefore, there is less upper-level divergence to intensify the cyclone, and the intensification rate slows over the next 12–24 h (not shown).

The evolution of the 1993 Superstorm demonstrates that:

- The greater the upper-level divergence is above a surface cyclone, the more the cyclone will intensify. In cases such as the 1993 Superstorm where a divergent region of a jet streak and strong CVA are colocated, a surface cyclone will intensify rapidly.

- The tilt of the 500-hPa trough is important. The more negatively tilted a 500-hPa trough is, the more concentrated CVA will be ahead of the trough (above the surface cyclone). This is because there is a shorter distance between the trough and the downstream ridge than with a positively tilted trough. All other things equal, surface cyclones intensify more when located downstream of a negatively tilted 500-hPa trough than they do when located downstream of a positively tilted 500-hPa trough.

- Surface cyclones will intensify when a 500-hPa trough axis is upstream of the surface cyclone. That is, in order for CVA to be located above the surface cyclone, the entire system must tilt westward with height. Similarly, a 500-hPa ridge axis must be located upstream of a surface anticyclone in order for AVA to be located above the surface anticyclone and for the surface anticyclone to intensify.

- 500-hPa troughs and ridges tend to move more quickly than surface cyclones and anticyclones, because winds in the upper-troposphere are faster than surface winds. As a result, 500-hPa troughs and ridges eventually catch up to surface cyclones and anticyclones, respectively. When this occurs, the intensification process slows and eventually stops, and the surface cyclone (anticyclone) achieves its minimum (maximum) pressure.

9.4 Qualitative Examination of Baroclinic Instability and Cyclone Lifecycle

In order to finalize a three-dimensional conceptual understanding of the evolution of mid latitude weather systems, we now examine the 1993 Superstorm at 500 and 850 hPa (Fig. 9.11). This allows us to visualize how the upper- and lower-troposphere interact with each other through a process called baroclinic instability.

At 00Z 13 March, the 500-hPa trough associated with the Superstorm is neutrally tilted and located over the western Gulf of Mexico (Fig. 9.11A). At the same time, a developing surface (lower-tropospheric) cyclone is located just northeast of the 500-hPa trough axis (Fig. 9.11E). The surface cyclone is located in a region of CVA, suggesting it will intensify. Behind the surface cyclone and directly underneath the 500-hPa trough is an area of strong CAA, which reduces the column thickness, resulting in lower 500-hPa heights and a stronger 500-hPa trough. Twelve hours later, both the 500-hPa trough and the surface cyclone have intensified rapidly (Fig. 9.11B and F).

The westward tilt with height of the entire system allows the surface cyclone to remain in an area of strong CVA, while strong CAA remains underneath the 500-hPa trough. This process, in which upper- and lower-tropospheric systems act to strengthen each other is called *baroclinic instability*, and is observed during the intensification of most midlatitude weather systems. Similarly, an upper-tropospheric ridge located upstream (west) of a lower-tropospheric anticyclone would cause both systems to intensify.

If baroclinic instability were to continue for an infinite period of time, both the upper- and lower- tropospheric would never stop intensifying. In the real atmosphere, however, baroclinic instability is a self-limiting process, because upper-tropospheric systems move more quickly than their lower-tropospheric counterparts. Fig. 9.11C and G show that by 00Z 14 March, the 500-hPa trough has nearly caught up to the 850-hPa cyclone, and 12 h later (12Z 14 March) the two systems are vertically stacked. At this point, the lower-tropospheric cyclone is not in a region of CVA and will no longer intensify. Meanwhile, the 500-hPa trough axis is no longer located above a region of CAA, indicating that the 1000–500-hPa thickness and 500-hPa heights will no longer fall. Thus the trough will cease to intensify and the surface cyclone will start to decay slowly due to surface friction.

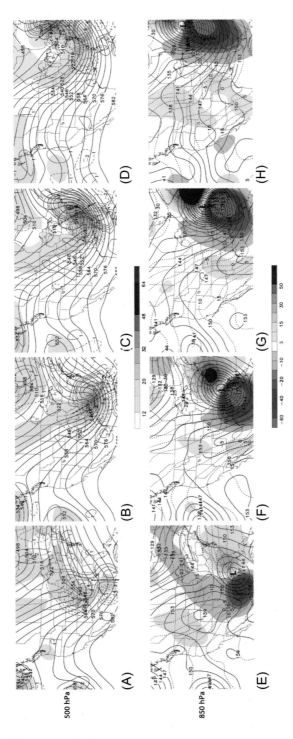

Fig. 9.11 (A–D) 500-hPa geopotential height (dam, *solid black contours*), geostrophic absolute vorticity (×10⁻⁵ s⁻¹, *shaded*), and geostrophic absolute vorticity advection (×10⁻⁹ s⁻², *blue contours, solid* for CVA, *dashed* for AVA), (E–H) 850-hPa geopotential height (dam, *solid black contours*), temperature (°C, *dashed black contours*), and geostrophic horizontal temperature advection (×10⁻⁵ K s⁻¹, *shaded warm colors* for WAA, *cool colors* for CAA). In each panel in the top (bottom) row, 500-hPa trough axes (lower-tropospheric cyclones) are marked with *dashed blue lines* (*red "L"*). Dates shown are (left) 00Z 13 March 1993, (second from left) 12Z 13 March 1993, (second from right) 00Z 14 March 1993, and (right) 12Z 14 March 1993.

9.5 Questions and Practice Exercises

1. You are making a forecast for Winnipeg, Manitoba, and observe the following:
   - Left entrance region of a straight jet streak.
   - 500-hPa AVA.
   - 850-hPa WAA.

   Discuss what you would expect the weather to be like in Winnipeg. You must explain your answers understand how each physical mechanism is associated with vertical motion.

2. CAA underneath a 500-hPa ridge axis will act to _____ the ridge. Explain your answer by drawing a schematic diagram.

3. True or false: Strong CVA ahead of a 500-hPa trough will cause the trough to move more quickly than weak CVA would.

4. Using the concepts of convergence and divergence, discuss whether a surface anticyclone located *upstream* (to the west of) of a 500-hPa ridge will intensify or decay.

5. A 988-hPa surface cyclone is located over central Minnesota. There is an area of strong WAA to the north of the cyclone and an area of strong CAA to the south. Where would you expect the surface cyclone to move over the next 12–24 h? Explain your answer.

6. A surface anticyclone is located underneath 500-hPa AVA and the right exit region of a jet streak. Will the surface anticyclone intensify or decay? Explain your answer.

7. Baroclinic instability states that a midlatitude weather system that tilts *eastward* with height will _____ over time. Explain your answer.

10

FRONTS AND DRYLINES

10.1 Fronts

10.1.1 Definition

A front is defined as "the interface or transition zone between two air masses of different density."[1] Although fronts are typically drawn on weather charts as lines, it is more useful to think of them as "zones." While there can be some immediate and/or dramatic changes between air masses as frontal "lines" suggest, the complete transition between air masses across a frontal zone is typically gradual.

Fronts occur most frequently in the midlatitudes, where different air masses collide. For example, in the winter, cold Arctic air ("Continental Polar" and "Continental Arctic") will move south and southeastward from the polar regions of Canada and northern Europe, eventually displacing warmer, moister ("Maritime Tropical") air originating in the subtropics. In spring, warm and moist air will move poleward, replacing the cool dry air that is in place throughout much of the winter. A few locations, most prominently the US Great Plains, experience the collision of three air masses in spring: Continental Polar air from the north; Maritime Tropical air from the Gulf of Mexico; and hot, dry air ("Continental Tropical") from the desert of the southwest United States and Mexican plateau. The collision of these three air masses along surface boundaries such as fronts and drylines (Section 10.2) sets the stage for severe weather, thus the nickname "tornado alley."

In this chapter, we discuss the characteristics of each type of front as well as drylines, and learn the best practices through which to identify each. We will also incorporate frontal analysis into sea-level pressure (SLP) charts, as detailed in Chapter 6.

---

[1] American Meteorological Society (AMS), 2017: Glossary of Meteorology, available online at: http://glossary.ametsoc.org/wiki/Front.

Synoptic Analysis and Forecasting. https://doi.org/10.1016/B978-0-12-809247-7.00010-7

10.1.2 Types

Fig. 10.1A shows the various types of fronts, while Fig. 10.1B shows other common surface boundaries and their respective symbols. Cold fronts are marked by blue lines with blue triangles pointing in the direction of frontal motion, while warm fronts feature red lines and red semicircles also oriented in the direction of motion. Stationary fronts are identified by alternating blue triangles and red semicircles on opposite sides of the alternating blue and red line. Finally, occluded fronts have alternating triangles and semicircles but are always purple. We will also discuss other common, nonfrontal surface boundaries in this chapter, specifically surface pressure troughs, squall lines, and drylines, indicated by the colors and symbols in Fig. 10.1B.

Fig. 10.2A shows schematics of a cold front in both the Northern and Southern Hemisphere (NH and SH). Recall that the source of cold air is the polar region, which is located to the north in the NH and the south in the SH. Think of the blue line (Fig. 10.2A) as the leading edge of the colder temperatures, which become gradually colder as we move farther behind the front. The blue triangles will always point in the direction that the front is moving or toward the warmer air that the cold air is replacing. In addition to being colder, the air behind the front is generally drier. As we will see in the next section, winds typically shift from a warm direction (e.g., south in the NH) ahead of the cold front to a cold direction (e.g., northwest in the NH) behind the cold front. Fig. 10.2B shows an x-z cross section of a cold front; think of the horizontal axis as west-east, while the vertical axis represents altitude. Recall that cold air is always more dense than warm air; therefore most cold fronts slope toward the colder air mass with height. In other words, most cold fronts pass through at the surface first, then later at progressively higher altitudes. As the surface cold air impinges on the warmer air mass ahead of the front, the warm air is forced to rise; rising motion, of course, frequently results in clouds and precipitation. The slope at a cold front tends to be much more steep than at a warm front, resulting in stronger ascent. This is one reason why thunderstorms, which require stronger updrafts, tend to occur more frequently with cold fronts than with warm fronts.

Fig. 10.3A shows schematics of a warm front for both the NH and SH. The source of warmer air is usually the tropics and subtropics, which are located to the south in the NH and the north in the SH. The red semicircles point in the direction that the warm front is moving or toward the colder air that the warm air is replacing. Air behind a warm front typically has a larger moisture

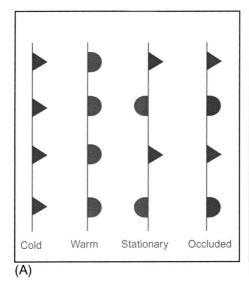 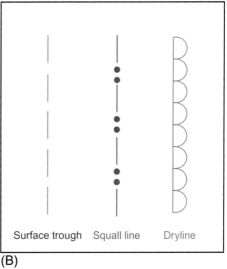

Cold   Warm   Stationary   Occluded   Surface trough   Squall line   Dryline

(A)   (B)

Fig. 10.1 Types of (A) fronts and (B) other boundaries plotted on sea-level pressure (SLP)/surface charts, using standard colors and symbols.

content than the air ahead of the front. Winds shift from a cool direction (e.g., northeast in the NH) ahead of the cold front to a warm direction (e.g., southwest in the NH) behind the warm front. In Fig. 10.3B, we see that the warm front passes at higher altitudes first, and at the surface last; that is, the warm front slopes toward the cooler air mass ahead of the front. In addition, the slope of the warm air rising above the cold air is gentler than near a cold front; this results in weaker ascent and less convective precipitation.

The third type of front is the stationary front (Fig. 10.4), drawn as an alternating blue and red line, with blue triangles pointing toward the warm air and red semicircles pointing toward the cold air. In the NH, cold air is generally located on the northern side of a stationary front, with warm air located to the south. Befitting their name, stationary fronts do not move; if and when they begin to travel, they become cold or warm fronts, depending on which air mass is advancing. The lack of motion of stationary fronts is best thought of in terms of the wind direction on either side. Recall that fronts move when one air mass advances on another; with stationary fronts, the wind on either side is opposite, but parallel to the front (Fig. 10.4B), not "pushing" the front to move in any direction. In the NH, the winds to the north of the front are typically from the east or northeast, while the winds south of the front are westerly or southwesterly. As we will see in Section 10.3, stationary fronts are the only

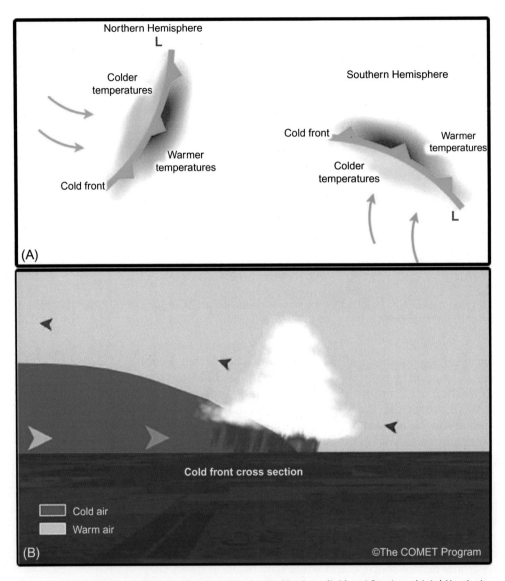

Fig. 10.2 For a cold front: (A) two-dimensional schematics for the Northern (left) and Southern (right) Hemisphere, with the typical location of a surface cyclone marked with a "L" and (B) an x-z cross section. Courtesy of the COMET website at: http://meted.ucar.edu/ of the University Corporation for Atmospheric Research (UCAR), sponsored in part through cooperative agreement(s) with the National Oceanic and Atmospheric Administration (NOAA), US Department of Commerce (DOC). ©1997–2016 University Corporation for Atmospheric Research. All Rights Reserved.

type of fronts not attached to a surface cyclone. In fact, there is usually a surface high pressure system on either side of the front, resulting in the parallel winds on either side. Stationary fronts can still be a focus for clouds and precipitation, however, which can occur on either or both sides of the front.

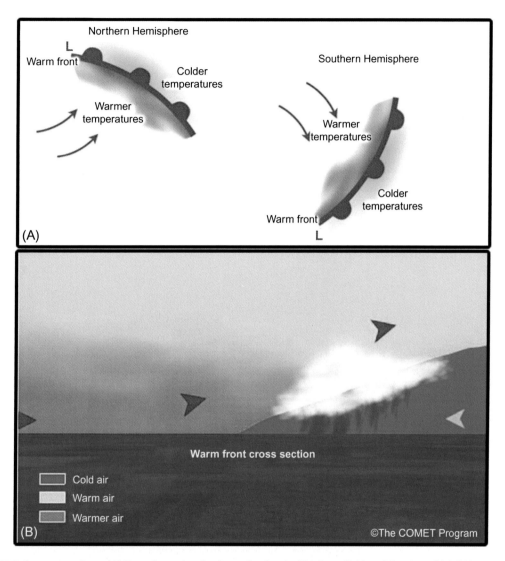

Fig. 10.3 For a warm front: (A) Two-dimensional schematics for the Northern (left) and Southern (right) Hemisphere, with the typical location of a surface cyclone marked with a "L" and (B) an x-z cross section. Courtesy of the COMET website at: http://meted.ucar.edu/ of the University Corporation for Atmospheric Research (UCAR), sponsored in part through cooperative agreement(s) with the National Oceanic and Atmospheric Administration (NOAA), US Department of Commerce (DOC). ©1997–2016 University Corporation for Atmospheric Research. All Rights Reserved.

The last front type discussed in this section is occluded fronts, shown in Fig. 10.5. Occluded fronts are marked with alternating triangles and semicircles pointed toward the direction of the front's motion and are drawn in purple. There are two types of occluded fronts (sometimes called "occlusions"): cold and warm. Cold occluded fronts, shown in the schematic in Fig. 10.5A, are the most

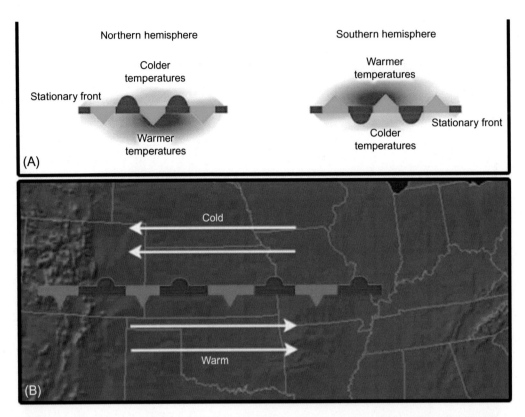

Fig. 10.4 For a stationary front: (A) two-dimensional schematics for the Northern (left) and Southern (right) Hemisphere, with the typical location of a surface cyclone marked with a "L" and (B) schematic showing winds parallel to the front on either side. Courtesy of the COMET website at: http://meted.ucar.edu/ of the University Corporation for Atmospheric Research (UCAR), sponsored in part through cooperative agreement(s) with the National Oceanic and Atmospheric Administration (NOAA), US Department of Commerce (DOC). ©1997–2016 University Corporation for Atmospheric Research. All Rights Reserved.

common type of occlusion; these fronts occur when a cold front, which typically moves faster than a warm front, catches up to a warm front. The warm, moist air mass between the cold and warm fronts shrinks in size, and eventually, the cold air behind the cold occlusion replaces the cool air ahead of the (former) warm front. Cold occlusions act a lot like cold fronts, in terms of wind shift, with a steep slope (Fig. 10.5B), and colder, drier air behind the front. However, the reductions in temperature and dew point across a cold occlusion are less dramatic than across a cold front. For example, imagine the air ahead of a cold front is 30°C (86°F) with a dew point of 20°C (68°F), while the air behind the cold front is 15°C (59°F) with a dew point of 5°C (41°F); that is a temperature and dew point difference of 15°C. Across a cold occluded front, the temperature and dew point differences would probably be closer to 5°C.

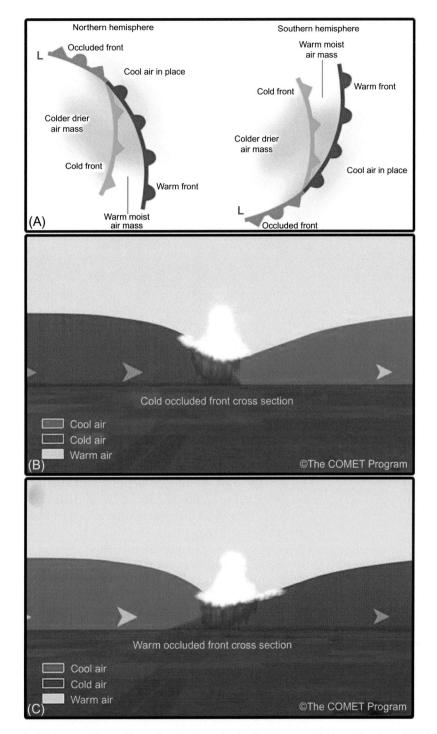

Fig. 10.5 For occluded fronts: (A) two-dimensional schematics for the Northern (left) and Southern (right) Hemisphere, with the typical location of a surface cyclone marked with a "L", with x-z cross sections for (B) a cold occluded front and (C) a warm occluded front. Courtesy of the COMET website at: http://meted.ucar.edu/ of the University Corporation for Atmospheric Research (UCAR), sponsored in part through cooperative agreement(s) with the National Oceanic and Atmospheric Administration (NOAA), US Department of Commerce (DOC). ©1997–2016 University Corporation for Atmospheric Research. All Rights Reserved.

The less common type of occlusion is a warm occluded front. As the cross section in Fig. 10.5C depicts, warm occlusions act similarly to warm fronts, but with a less drastic temperature and dew point contrast. The temperature of the air behind a warm occluded front is still cool, but not as cold as temperatures ahead of the front. Warm occlusions are rare because they are limited to certain regions; in order for air behind a surface cyclone to be warmer than the air ahead of it, the source region has to be unique. One of the most common regions for warm occlusions is the northwest coast of the United States and the southwestern coast of Canada. Although the Pacific Ocean water temperatures are not warm (∼5–10°C, 40–50°F), in the winter this "Maritime Polar" air is frequently warmer than the Continental Polar air in the mountainous interior. It requires a bit of counter-intuitive thinking, but as the front passes, relatively mild Pacific air from the west replaces cold continental air located to the east. Warm occlusions tend to occur only in the cold season, in locations where a relatively mild ocean is upstream (upwind) of a colder continent.

10.1.3 Identification

Although all types of fronts *can* be associated with precipitation, they do not have to be. Fronts can frequently move across a given region without any precipitation. As a result, the analyst or forecaster should not rely on the presence (or lack thereof) of precipitation to diagnose a front. Instead, the following four criteria should be used to determine the type of front and its location:

- Temperature change across the front.
- Dew point change across the front.
- Wind shift across the front.
- SLP minimum along the front; SLP falls as the front approaches, and rises after it passes.

Fig. 10.6 shows an example of a central US cold front, marked by the line with the triangles. The cold front is moving from the northwest toward the southeast. In Fig. 10.6A, we see that the temperature (isotherms, red dashed lines) gradually decreases from >70°F (21°C) ahead of the front to <60°F (16°C) well behind the front. While there is a temperature change immediately in the vicinity of the front, the temperature continues to fall well behind the front, emphasizing that fronts are truly zones, not thin lines. The green circles in Fig. 10.6A highlight the dew point decrease from ahead of the front to behind it; ahead of the front, dew points are 61–64°F (16–18°C), but fall to approximately 59°F (15°C)

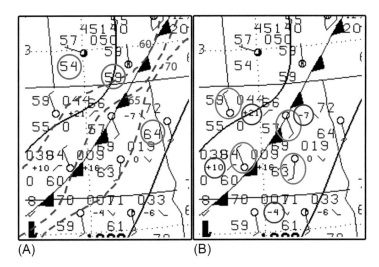

Fig. 10.6 Surface observations across a cold front (*bold line with triangles*) highlighting the (A) temperature (°F, *red dashed*) and dew point (°F, *green circles*) contrasts, and (B) wind shift (*kt, blue circles*) and SLP tendency (tenths of a hPa, *purple circles*).

immediately behind the front, and 54°F (12°C) farther behind the front. This emphasizes that air masses behind cold fronts are generally drier as well as colder.

Fig. 10.6B focuses on the wind shift and SLP tendency change across the cold front. The blue circles highlight that the wind ahead of the cold front is primarily from the south and southeast, both warm wind directions. Behind the cold front, winds are from the northwest, which is a cold direction. Another way to think of it is that the wind rotates cyclonically (counterclockwise in the NH) across the front. The purple circles in Fig. 10.6B indicate 3-h SLP tendencies, following the surface station plot model introduced in Chapter 2. All fronts, no matter which type, are SLP minima. Therefore SLP tendencies ahead of the cold front in Fig. 10.6B are negative, indicating that SLP is falling as the front approaches. Behind the front, SLP is rising, as evidenced by the positive SLP tendencies.

Table 10.1 shows the characteristics of temperature, dew point, wind shift, and SLP tendency respective to each type of front, as well as where—relative to each front type—one can expect precipitation if it occurs. Recall that cold occluded fronts act like cold fronts, with less drastic temperature and dew point changes across the front, while warm occlusions similarly mimic warm fronts. All fronts are SLP minima, including stationary fronts.

Table 10.1 Characteristics of Each of the Five Front Types

Front Type	Temperature	Dewpoint	Wind Shift	SLP	Precipitation
Cold	Warmer ahead, colder behind	Moister ahead, drier behind	Warm direction ahead, cold direction behind	Falling ahead, rising behind	Ahead, at, and/or behind
Warm	Colder ahead, warmer behind	Drier ahead, moister behind	Cold direction ahead, warm direction behind	Falling ahead, rising behind	Ahead and/or along
Stationary	Colder poleward, warmer equatorward	Drier poleward, moister equatorward	Parallel but opposite on either side of the front	Minimum along front	Along and/or on either side
Cold occluded	Warmer ahead, colder behind	Moister ahead, drier behind	Cool direction ahead, cold direction behind	Falling ahead, rising behind	Ahead, at, and/or behind
Warm occluded	Colder ahead, warmer behind	Drier ahead, moister behind	Cold direction ahead, cool direction behind	Falling ahead, rising behind	Ahead and/or along

SLP will increase as you move farther away from the front. With warm fronts, however, the SLP rise after the front passes can sometimes be obscured by the fact that the cold front (and corresponding falling SLP) typically approaches in short order thereafter.

The presence of precipitation should typically not be used to analyze whether a front exists or to discern between front types. However, Table 10.1 details where precipitation would exist with respect to each front type, if it occurs. For warm fronts, precipitation is typically located at or ahead of the front; this is because warm air rises gently over the colder air ahead of the front, causing clouds and precipitation ahead of the front. In contrast, precipitation can occur anywhere relative to a cold front, including behind it. Sometimes there may be a prefrontal squall line, marked with the red line and symbols in Fig. 10.1B, followed by an area of lighter or stratiform precipitation along and/or behind the front. Precipitation can be along and/or on either side of a stationary front, while precipitation with cold and warm occlusions is generally similar to precipitation associated with cold and warm fronts, respectively.

Finally, a good rule of thumb for frontal identification is to examine whether any three of the first four criteria listed in Table 10.1 (temperature change, dew point change, wind shift, SLP tendency change) are satisfied. Through knowledge of the characteristics of each front type, as well as practice and repetition, the analyst or forecaster can regularly produce an accurate frontal analysis. One should be mindful, however, that frontal analysis is relatively subjective. Even if you gathered 20 experienced meteorologists in the same room to perform the same frontal analysis, each plot would be slightly different. However, manual frontal analysis performed by an analyst or forecaster is still much more skillful than an automated computer frontal analysis; computers have advanced meteorology in many ways, but automated frontal analysis still lags behind in accuracy.

10.2 Drylines

10.2.1 Definition

The symbol for a dryline is shown in Fig. 10.1B; while not technically a front, a dryline is an important type of surface boundary that can serve as a focus for clouds, precipitation, and severe thunderstorms. One can think of a dryline as a "dew point front," with similar dew point and wind changes to a cold front, but no temperature decrease. Drylines occur when hot Continental Tropical air meets warm and moist Maritime Tropical air.

Fig. 10.7 shows horizontal and cross section schematics of a dryline in the US Great Plains, which is by far the most frequent region of occurrence. Other locations that see occasional drylines are the Mediterranean region of Europe (e.g., Greece, Italy) and Australia. In Fig. 10.7A, hot, dry Continental Tropical air from the US desert southwest and Mexican plateau is impinging on warm, moist Maritime Tropical air from the Gulf of Mexico. Dew points across a dryline can drop by as much as 60–70°F (20–35°C). While dew points change drastically across a dryline, temperature does not. In fact, temperature is frequently warmer behind a dryline than ahead of it, because of fewer clouds (drier air) in that region. Similar to fronts, there is a wind shift across a dryline (Fig. 10.7A), with winds converging at the dryline.

Recall that moist air is less dense than dry air, because of its larger concentration of water vapor. As a result of the converging winds and the density difference, air will be forced to rise at the dryline. If there is enough ascent, moisture, and instability, thunderstorms will result. Drylines are a key thunderstorm formation

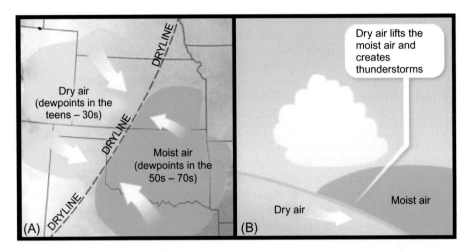

Fig. 10.7 For a dryline: (A) two-dimensional schematic over the Great Plains of the United States, and (B) an x-z cross section. Courtesy of the COMET website at: http://meted.ucar.edu/ of the University Corporation for Atmospheric Research (UCAR), sponsored in part through cooperative agreement(s) with the National Oceanic and Atmospheric Administration (NOAA), US Department of Commerce (DOC). ©1997–2016 University Corporation for Atmospheric Research. All Rights Reserved.

location in the US Great Plains ("tornado alley"), particularly in spring and early summer when the Mexican plateau is most heated. As upper-tropospheric troughs and jet streaks provide large-scale ascent, drylines often provide a mesoscale ascent-focusing mechanism for severe storms to form. In the next section, we will see an example of how a dryline can be associated with severe weather and tornado formation.

10.2.2 Identification

Fig. 10.8 shows surface station plots and a manually analyzed dryline at 18Z on 20 May 2013, just before the Moore, Oklahoma EF-5 tornado. The dryline is oriented from northeast to southwest, with the northern end located just west of the Oklahoma City metro area. Note that the temperature and dew point at Oklahoma City are 82°F (28°C) and 71°F (22°C), respectively, while at Amarillo, Texas, located 418 km (260 miles) to the west, the temperature is 84°F (29°C), but the dew point is only 21°F (−6°C). This is a 50°F (28°C) dew point drop across the dryline! Similar dew point differences are seen from east to west over the length of the dryline. While temperatures are similar on either side of the dryline, there is a stark wind shift from south-southeasterly ahead of the dryline to west and west-northwest behind the dryline.

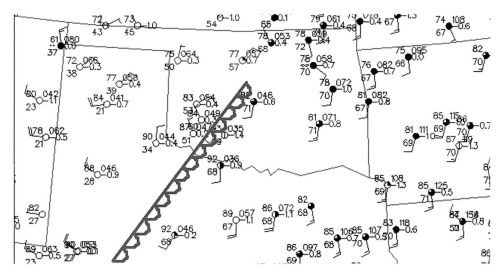

Fig. 10.8 Surface station plots and dryline (*brown-hashed line*) over Oklahoma and Texas at 18Z 20 May 2013, just before the Moore, Oklahoma EF-5 tornado. From the Plymouth State University Weather Center Make Your Own: Surface Data, available online at: https://vortex.plymouth.edu/myo/sfc/.

This enables surface convergence at the dryline, enhancing ascent and increasing the chance of thunderstorms.

As we will see in Chapter 14, the environment in the Oklahoma City (Moore) area that day was characterized by strong instability, large amounts of moisture (Fig. 10.8), and divergence aloft (i.e., large-scale ascent). The dryline in Fig. 10.8 increased surface convergence (i.e., provided additional ascent) and served as an air mass (density) boundary on which thunderstorms could initiate. The clash of air masses evidenced in Fig. 10.8 exemplifies why the US Great Plains experiences the most severe thunderstorms and tornadoes in the world.

10.3 Surface Analyses and Practice Examples

10.3.1 Surface Analyses

Now that we have discussed all fronts and drylines, we can put the entire picture together by examining the North American surface analysis in Fig. 10.9. There are four notable surface cyclones across North America: one off the California coast, one in

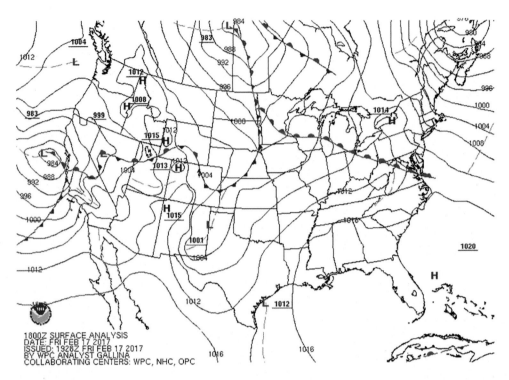

Fig. 10.9 National Oceanic and Atmospheric Administration (NOAA) National Weather Service (NWS) Weather Prediction Center (WPC) North American surface analysis with fronts, from 18Z 17 February 2017. SLP (hPa, *solid brown lines*), surface cyclones and anticyclone centers (*red* "L" and *blue* "H," respectively), and fronts are plotted. From NOAA NWS WPC: North American Surface Analysis, available online at: http://www.wpc.ncep.noaa.gov/html/sfc2.shtml.

northern Manitoba, a third in the Texas Panhandle, and the final one just off the map in Atlantic Canada. Both the cyclone off the California coast and the one in northern Manitoba feature an occluded front located immediately equatorward of the cyclone center, with warm and cold fronts located farther equatorward. The presence of an occluded front indicates that each cyclone is past its peak intensity stage and starting to weaken. The occluded front attached to the Manitoba cyclone is a cold occlusion, but the occluded front attached to the California cyclone is a warm occlusion, as mild Pacific winter air encroaches on colder temperatures over the continent.

As a rule, occluded, cold, and warm fronts must connect to a surface cyclone center even if, as in Fig. 10.9, the cold and warm fronts first connect to the occluded front. Cold, warm, and occluded fronts should never be connected to a surface high pressure system. The only type of front that can be drawn on a frontal analysis without being connected to a surface cyclone is a stationary front. There are two stationary fronts in Fig. 10.9, one across Nevada and Utah, and the other draped across northern Manitoba and northwestern Ontario. Finally, note the thin orange dashed lines in several locations on the surface analysis, including near the surface cyclone in the Texas panhandle; as shown in Fig. 10.1B, these dashed lines are indicative of surface pressure troughs. Surface pressure troughs represent axes of minimum SLP and wind shifts; however, unlike fronts, the temperature and dew point contrasts across the surface pressure troughs are not substantial.

10.3.2 Practice Examples

Fig. 10.10 features two charts containing surface station plots, with manually analyzed fronts using the appropriate symbols. Fig. 10.10A shows a cold front oriented from northeast-to-southwest across Illinois and Missouri; the surface cyclone to which the cold front is attached is located over northern Lake Michigan off the map, and is therefore not pictured. To the east of the cold front, temperatures are 75–85°F (23–29°C) and dew points are >70°F (21°C). The wind directions ahead of the cold front are generally from the south-southwest, switching to northwest behind the front. Temperature and dew point begin to gradually decrease just behind the front and continue to drop markedly into Iowa, Nebraska, and southern Minnesota (Fig. 10.10A). In this particular case, precipitation is primarily

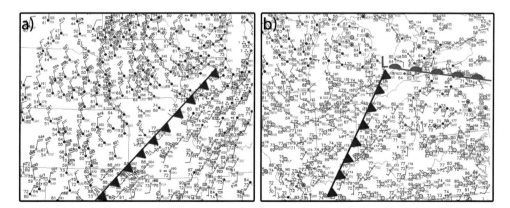

Fig. 10.10 (A,B) Two maps of surface station plots with analyzed fronts and surface cyclone, as appropriate, using typical plotting conventions. In (A), the surface cyclone center is located off the map (to the northeast of the cold front), so an "L" is not plotted. Modified from the National Center for Atmospheric Research (NCAR) Research Applications Laboratory (RAL): Real-time weather data, surface, available online at: http://weather.ral.ucar.edu/surface/.

located just ahead of the front, with a few thunderstorms along the front; skies are generally cloudy on both sides of the cold front.

Fig. 10.10B shows a surface cyclone centered over southern Lake Huron, as evidenced by the cyclonic rotation in the wind around the low pressure center. A warm front extends east-southeast of the cyclone center and a cold front is located to the south-southwest. The air to the north of the warm front is relatively cool and dry compared to the warm, moist air over Pennsylvania and Ohio, in the warm sector between the two fronts. Wind directions ahead of the warm front are mostly from the east, a relatively cool direction, but from the south in the warm sector, advecting warm and moist air northward. Across the cold front, temperature, dew point, and wind direction changes are similar to the cold front in Fig. 10.10A, but with a less drastic temperature change. Some frontal analysts may have chosen to draw a cold occluded front instead of the cold front in Fig. 10.10B, emphasizing the subjectivity associated with frontal analysis. Finally, notice the lack of precipitation near either front; as mentioned earlier in this chapter, precipitation should not be used as a frontal discriminator.

10.4   Questions and Practice Exercises

1. Draw a northeast-to-southwest oriented warm occluded front, using the appropriate symbols and plotting conventions.
2. Which is the only type of front that does not have to be connected to a surface cyclone on a frontal analysis/surface chart?
3. Name the four specific criteria you would look for in order to draw a warm front on a surface analysis.
4. Why should precipitation not be used as a frontal discriminator?
5. Describe the wind direction on either side of a stationary front.
6. A SH cold front is moving toward the northeast. Which direction is the cold air coming from?
7. Behind a dryline, dew points will (increase/decrease/stay the same), while temperatures will (increase/decrease/stay the same).
8. Examine the surface station plot in Fig. 10.11. Using appropriate plotting conventions, draw:
   - The surface cyclone center.
   - Any fronts that you see.

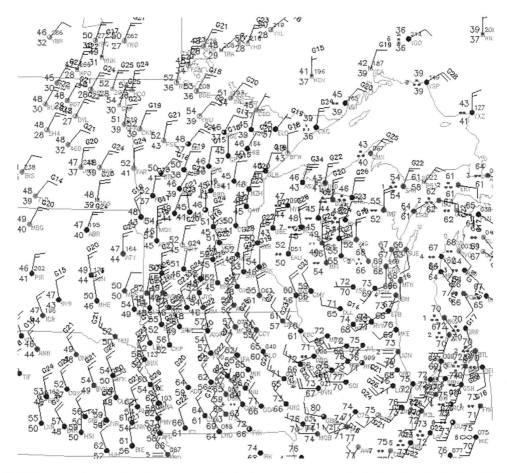

Fig. 10.11 Map of surface station plots to be used for the frontal analysis exercise in Question. Modified from NCAR RAL: Real-time weather data, surface, available online at: http://weather.ral.ucar.edu/surface/.

11

SATELLITE IMAGERY

11.1 Introduction to Weather Satellites

11.1.1 How They Work

When the first weather satellite was launched in 1960,[1] it revolutionized weather observations and forecasts. Along with the development of the first weather instruments (e.g., thermometer, barometer), Doppler Radar, and increased computing power, satellite technology is one of a handful of technological advances that has had an enormous impact in meteorology. By the time a global network of weather satellites was established in the 1970s, meteorologists were, for the first time, able to track weather systems throughout the atmosphere in real time. Previously, weather observations were limited to the Earth's surface, with the exception of radiosondes and occasional aircraft data; this was a particular problem over data-sparse regions such as the oceans. As satellite technology has improved over time, satellite data are used throughout most sectors of atmospheric science, including weather analysis and forecasting, air pollution, and climate science.

Weather satellites are passive remote sensors; unlike radars, they do not emit a beam of energy. Instead, they measure energy emanating from Earth's surface and atmosphere using different wavelengths, or channels, at which individual gases such as water vapor (WV) selectively emit and absorb radiation. Although satellites are used for a multitude of observational and research endeavors, in this chapter we will primarily focus on three types of satellite imagery used for analysis and forecasting: visible (VIS), infrared (IR), and WV.

---

[1] The first weather satellite was TIROS I, launched on 1 April 1960. Details about its launch, imagery, and capabilities can be found on the National Oceanic and Atmospheric Administration (NOAA) Satellite and Information Service website, available online at: http://noaasis.noaa.gov/NOAASIS/ml/40yearsa.html.

Synoptic Analysis and Forecasting. https://doi.org/10.1016/B978-0-12-809247-7.00011-9

11.1.2 Types of Weather Satellites

Before we explore different types of satellite imagery, it is important to understand the two types of weather satellites, polar-orbiting and geostationary, and how they work. We will spend most of our time discussing geostationary satellites, because they provide continuous coverage of a given region in real time. Fig. 11.1 shows a comparison of the two orbit types: polar-orbiting satellites orbit from pole to pole, carving out a relatively narrow longitude swath with each scan. These satellites complete a single longitude swath approximately every 100 min and typically will scan any point twice a day. Obviously, scanning a given point on Earth only twice a day is limiting, because many weather features such as thunderstorms, tornadoes, and even fair-weather clouds change on much smaller timescales (minutes to an hour). However, polar-orbiting satellites do possess several advantages that help atmospheric scientists monitor the Earth system. First, they are much closer to Earth's surface than geostationary satellites; polar-orbiting satellites are approximately 850 km above the surface, compared to 36,000 km for geostationary satellites. A lower orbit allows polar-orbiting satellites to resolve atmospheric features with a finer resolution; however,

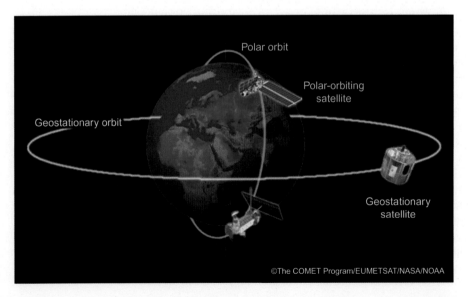

Fig. 11.1 Schematic of the typical orbits for polar-orbiting and geostationary satellites. Courtesy of the COMET website at: http://meted.ucar.edu/ of the University Corporation for Atmospheric Research (UCAR), sponsored in part through cooperative agreement(s) with the National Oceanic and Atmospheric Administration (NOAA), US Department of Commerce (DOC). ©1997–2016 University Corporation for Atmospheric Research. All Rights Reserved.

due to technological advances in the latest geostationary satellites, this gap is closing. Most importantly, polar-orbiting satellites are the only source of satellite data for the polar regions (poleward of approximately 70° in both hemispheres). They provide crucial information to help monitor arctic weather and climate, including ice sheets, sea ice, and polar lows.

Geostationary satellites are the primary source of satellite imagery for real-time weather monitoring. Each geostationary satellite orbits approximately 36,000 km above the Earth's surface, usually near the equator, and moves in perfect sync with the Earth's rotation. The high orbit allows each geostationary satellite to have continuous imagery from the same geographical domain 24 h a day, 365 days a year. Typically, geostationary satellites cannot see poleward of 70° latitude in either hemisphere, although this can sometimes vary depending on a satellite's position. The imagery emanating from geostationary satellites has traditionally been of substantially lower resolution than polar-orbiting imagery, although that is changing somewhat with the latest generation of satellites.

Fig. 11.2 illustrates many of the current geostationary and polar-orbiting satellites, as of early 2017. Satellites are owned

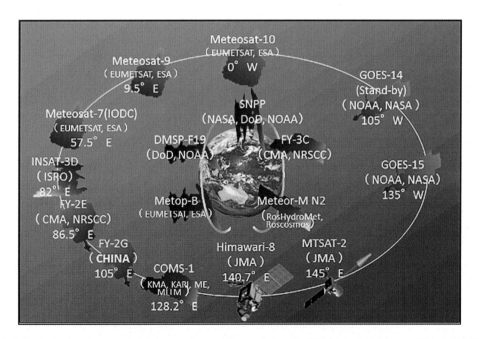

Fig. 11.2 Schematic showing most current weather satellites as of early 2017. Note that the new US satellite GOES-16 (East), typically positioned at 75°W, is not included in the schematic. From the Japan Meteorological Agency (JMA): Global Meteorological Observation, available online at: http://www.jma-net.go.jp/msc/en/general/system/www/index.html.

and maintained by various countries and organizations across the globe. For example, the European Space Agency operates three geostationary satellites (Meteosat-7, -9, and -10) that cover portions of the Atlantic Ocean, as well as all of Europe, Africa, and the Middle East. In the western Pacific, the Japan Meteorological Agency operates Himawari-8, a next-generation satellite that has revolutionized how we observe the evolution of western Pacific tropical cyclones. In the western Hemisphere, the primary satellites are the Geostationary Operational Environmental Satellites (GOES), owned and operated by the US National Aeronautics and Space Administration (NASA). Currently, there are three GOES in orbit: GOES-15 (GOES-West, western North America and the Eastern Pacific), GOES-14 (a reserve satellite), and the recently launched GOES-16 (GOES-East, eastern North America, Central and South America, and the western Atlantic). GOES-16, which replaced GOES-13, is the first of the new generation of NASA satellites; the next-generation replacement for GOES-West, tentatively called "GOES-S," is scheduled to be launched in 2018.

The global geostationary satellite network has enough overlap such that if one satellite fails, adjacent satellites can temporarily fill the void. Fig. 11.3 shows simultaneous full-disk IR images from Himawari-8 (western Pacific) and GOES-15 (eastern Pacific). Notice that the area of clouds (bright white) in the equatorial Central Pacific can be seen on both images. In addition to providing a failsafe in the case of an equipment issue, the overlap between geostationary sectors allows meteorologists to seamlessly and

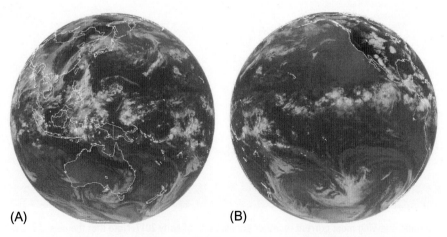

(A) (B)

Fig. 11.3 Full-disk infrared (IR) satellite images from (A) Himawari-8 and (B) GOES-West, taken at the same time on 3 July 2017. Data from NOAA: Geostationary Satellite Server, available online at: http://www.goes.noaa.gov/index.html.

continuously observe weather features as they move and evolve. In fact, many online providers of satellite data will use computer algorithms to stitch together images from multiple geostationary satellites, in order to provide the best possible imagery for the region of interest. For example, a North American domain view would be a compilation of GOES-West and GOES-East, as shown in the schematic in Fig. 11.4.

11.2   Common Types of Satellite Imagery

In this section, we explore the three most common types, or bands, of satellite imagery used for weather analysis and forecasting: VIS, IR, and WV. Although modern satellites work at many more than three bands, our goal here is not to replicate a remote sensing course experience, but instead reinforce the utilities of satellite imagery for analysis and forecasting.

11.2.1   Visible

The first type of satellite imagery, VIS, is based upon the principle of reflected sunlight. The amount of reflected sunlight that the satellite can remotely sense is dependent on the reflectivity, or albedo, of the surface off which the sunlight is reflecting.

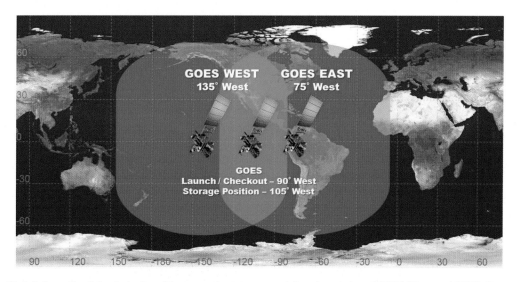

Fig. 11.4 Schematic of the typical positions, areal coverage, and domain overlap of GOES-West and GOES-East. From National Aeronautics and Space Administration (NASA): GOES, available online at: https://www.nasa.gov/sites/default/files/thumbnails/image/goes-r_fleet_new6.png.

Table 11.1 Reflectivity of Visible Sunlight for Common Surfaces, Including Cloud Types

Surface	Reflectivity (%)
Cumulonimbus (thunderstorm)	90
Fresh snow	85–90
Thick stratus or stratocumulus	70
Altostratus	40
Cirrus	25
Forest	12
Water	9

Table 11.1 presents reflectivities for a few common surfaces, including fresh snow cover (extremely reflective), water (not very reflective), and various cloud types. We can see that cumulonimbus (thunderstorm) clouds have the largest reflectivity among cloud types. This is because the amount of reflected sunlight is generally proportional to the thickness, or depth, of a cloud. On a VIS satellite image, the thickest clouds (e.g., cumulonimbus) appear the brightest white, while thin clouds (e.g., cirrus) appear dim gray. Land surfaces, especially bodies of water, have the smallest reflectivities and therefore appear dark.

Another important consideration for VIS imagery is that it only works during daylight hours. Fig. 11.5 shows two GOES-16 VIS images of the southwest United States and northern Mexico, with Fig. 11.5A taken at 13Z and Fig. 11.5B taken 3 h later, at 16Z. Recall that in the summer, 13Z is 6 a.m. Pacific Daylight Time, which is before sunrise in places such as Arizona and California. Therefore, in Fig. 11.5, we see clouds from parts of New Mexico and Colorado eastward, but everything to the west of there appears dark. This does not mean the satellite is broken! Instead, it indicates that there is no reflected sunlight for the satellite to sense. Three hours later, Fig. 11.5B shows cloud features throughout the entire domain, including stratus and fog just off the Pacific coast.

Fig. 11.6 shows three VIS images from the same time on 3 July 2017. In Fig. 11.6A, there are numerous thunderstorms (cumulonimbus) located over Florida and the eastern Gulf of Mexico. We know they are thunderstorms because of their cellular shape and bright white color, which indicates very thick clouds. If we

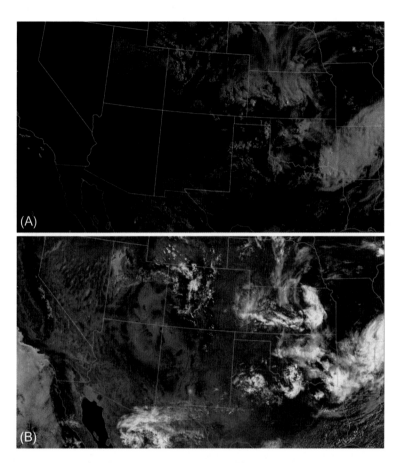

Fig. 11.5 Visible (VIS) satellite imagery from GOES-16 over the southwestern United States and northern Mexico on 3 July 2017 at (A) 13Z and (B) 16Z. Reproduced with permission from College of DuPage NEXLAB: GOES-16 imagery, available online at: http://weather.cod.edu/satrad/exper/.

examine Fig. 11.6A closely, we can see that the anvils on the edges of the cumulonimbus appear gray, because they are thinner. In Fig. 11.6B, there is a clear contrast between the thicker clouds located throughout most of Pennsylvania and the thin wispy gray features to the south over Maryland, Washington, District of Columbia, and Virginia. Notice that the clouds in Pennsylvania, while bright, are not quite as bright white as the thunderstorms over Florida (Fig. 11.6A), suggesting that they are less thick than cumulonimbus, likely towering cumulus or stratocumulus. The gray wispy clouds to the south are cirrus, and if we were to animate the image, we would see them racing off to the east with the jet stream.

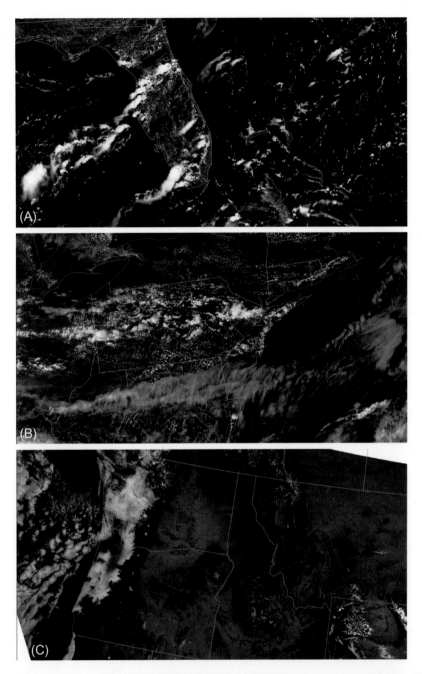

Fig. 11.6 VIS satellite imagery from GOES-16 at 18Z 3 July 2017 over (A) Florida and surrounding waters, (B) the US Mid-Atlantic states, and (C) the Pacific Northwest of the United States and far southwestern Canada. *Brighter white* indicates more reflective, and thus thicker clouds. Reproduced with permission from College of DuPage NEXLAB: GOES-16 imagery, available online at: http://weather.cod.edu/satrad/exper/.

Fig. 11.6C shows mostly clear skies over the interior Pacific Northwest United States. If you look closely, you might be able to identify specific river valleys and mountain ranges, especially in Washington, Oregon, and Idaho. Of particular interest are the bright gray features located along most of the Washington and Oregon coastlines. Note that these clouds are not as bright as the Florida thunderstorms or Pennsylvania stratocumulus, but still appear thicker than the cirrus over Maryland and Virginia (Fig. 11.6B). If we were to animate this image, we would see that the cloud features do not move, which indicates that they must either be fog/low stratus or snow cover. Given that the image is from July and snow cover would appear more reflective (brighter white), we can deduce that this is fog/low stratus, which is very common at night and early in the morning in the Interstate-5 corridor between the Canadian border and Los Angeles. We further discuss how to identify fog using multiple types of satellite imagery and other sources in Section 11.3.

11.2.2 Infrared

All objects with a temperature greater than absolute zero radiate energy. The wavelength of the emitted radiation is dependent on the temperature of the object. Because the Earth's temperature is relatively low (as compared to, e.g., the sun), it primarily radiates energy in the IR portion of the radiation spectrum. Weather satellites detect the emitted IR radiation from both the surface and the atmosphere. By measuring IR radiation, satellites can infer cloud top temperature, which in turn can be used to infer cloud top height. Therefore, whereas VIS satellite imagery can provide information about cloud thickness, IR imagery primarily provides cloud-top temperature and height information. Finally, since IR radiation is not dependent on visible light, IR satellite imagery can be used at any time of day.

Fig. 11.7 shows IR satellite images from the identical locations and times as the VIS images in Fig. 11.6. False color is added to the images to emphasize clouds with higher (colder) tops. The thunderstorms over Florida in Fig. 11.7A clearly have the coldest (highest) cloud tops; specifically, the orange and red in the thunderstorm over southwest Florida indicate a very tall cumulonimbus cloud. In Fig. 11.7B, the cirrus over Maryland, District of Columbia, and Virginia are shaded blue, green, and purple, indicating that they are high-topped clouds, but not as high as the cumulonimbus in Florida. This makes sense, as the average cumulonimbus cloud is 30,000–40,000 ft. tall, while cirrus clouds are typically located between 20,000 and 30,000 ft. Now examine

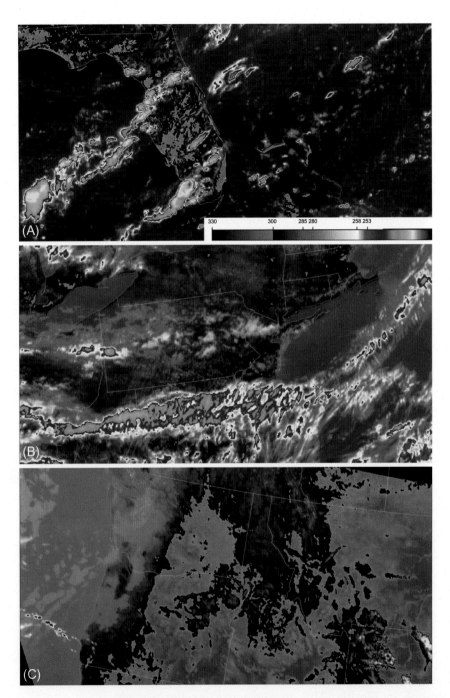

Fig. 11.7 As in Fig. 11.6, but for infrared (IR) satellite imagery. False color is added to the images to highlight higher-topped clouds, as shown in the color bar (K, *shaded*) at the bottom of panel (A); the highest (coldest) cloud tops are indicated in *orange* and *red*. Reproduced with permission from College of DuPage NEXLAB: GOES-16 imagery, available online at: http://weather.cod.edu/satrad/exper/.

the stratocumulus over Pennsylvania (Fig. 11.6B); with the exception of a higher-topped cloud in far western Pennsylvania, the tops are clearly lower than both the cirrus to the south and the cumulonimbus over Florida.

The lower in the atmosphere a cloud is, the warmer it is, and the darker gray it appears on an IR satellite image. For example, the temperatures of stratus and especially fog are very similar to the Earth's surface temperature. Examine the fog over Washington and Oregon that we identified in Fig. 11.6C; on the IR image (Fig. 11.7C), the fog is shaded dark gray, which is somewhat similar to the shading of the land surface. This is one of the drawbacks to exclusively using IR satellite imagery; it is difficult to spot cloud features close to the ground. The best course of action in these cases, when feasible, is to use multiple types of satellite imagery, as well as other weather data (e.g., surface observations).

As we have seen, VIS imagery should be used to assess cloud thickness, while IR imagery is best for cloud top temperature and height. To maximize their utilities, the analyst or forecaster should use them together when possible, i.e., during daytime hours. Table 11.2 shows common combinations of VIS and IR observations, and the likely associated cloud type. For example, when clouds appear bright on both VIS and IR, they must be thick with high cloud tops, likely cumulonimbus or towering cumulus. Cirrus clouds, which are high, will be bright on IR, but wispy and gray on VIS because they are thin. Low stratus and fog are located close to or at the ground, meaning they are dark gray (warm) on IR, but they also tend to be relatively thick, and thus appear fairly bright gray or dull white on VIS. Finally, fair-weather cumulus and stratocumulus are fairly thin and warm (low), so they will generally be gray on both VIS and IR. Although we cannot cover every

Table 11.2 Appearance Descriptions for Frequently Observed Clouds on Both VIS and IR Imagery

VIS Cloud Appearance	IR Cloud Appearance	Possible Cloud Type
Thick (bright white)	Cold (very bright)	Cumulonimbus, towering cumulus, or nimbostratus
Thin (dull gray)	Cold (bright)	Cirrus
Medium (dull white)	Warm (dull gray)	Low stratus, fog
Medium-thin (gray)	Warm (gray)	Cumulus, stratocumulus

cloud type in this text, it is important that the analyst or forecaster practice identifying important cloud types using a combination of VIS and IR imagery.

11.2.3 Water Vapor

The third and final type of satellite imagery discussed in this text is WV imagery. The WV channel is actually within the IR spectrum but is designed to detect atmospheric moisture. Because the WV channel is detecting moisture from space, WV imagery will show moisture closest to the satellite, which is typically in the upper troposphere (~200–500 hPa). If the upper troposphere is dry, the WV channel is then able to detect lower-tropospheric moisture.

There are several utilities of WV satellite imagery, but the simplest is its ability to discern dry and moist regions. Fig. 11.8 shows two GOES-16 WV images; on this particular false color scale, orange and red indicate very dry regions, while blue, green, and white indicate moist regions. In the full disk image centered on the equator (Fig. 11.8A), there is an area of large WV content just at and north of the equator, in association with the Inter-Tropical Convergence Zone (ITCZ). In contrast, within the tropical and subtropical Southern Hemisphere, there is a large region of extremely dry air. In Fig. 11.8B, taken at the same time as the VIS and IR images in Figs. 11.6A and 11.7A, respectively, the largest

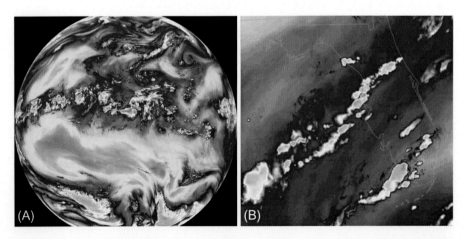

Fig. 11.8 GOES-16 Water Vapor (WV) satellite imagery at 18Z 3 July 2017 for (A) full disk and (B) Florida and surrounding waters. *Yellow, orange, and red areas* indicate dry air, while *blue, green, and white areas* are indicative of moist air. Reproduced with permission from College of DuPage NEXLAB: GOES-16 imagery, available online at: http://weather.cod.edu/satrad/exper/.

moisture content is associated with the thunderstorms over Florida. In Fig. 11.7A, we saw that the thunderstorm located in southwest Florida had the highest (coldest) cloud tops, and Fig. 11.8B shows that it also has the largest upper-tropospheric WV content. Finally, notice the contrast between the blue (moist) area across most of the Florida peninsula and adjacent waters, and the yellow/gold area over the Florida panhandle; this is indicative of a moisture gradient, with larger moisture content located to the south, and drier air to the northwest.

WV imagery can also be useful in diagnosing multiple types of atmospheric waves. For example, mountain and gravity waves can often be detected by looking for "ripple" features in the WV imagery. In this chapter, we will focus on how to utilize WV imagery to detect large-scale upper-tropospheric features such as troughs and ridges, discussed in the next section.

11.3 Feature Identification

In this section, we explore how to best utilize the three satellite imagery types introduced in Section 11.2 to diagnose important weather features. The weather features in this section are by no means a comprehensive list, and the reader is encouraged to seek additional satellite imagery user guides and learning modules. The new generation of satellites have the capability to readily detect small-scale features such as lightning strikes and fires, and thus may be used for many applications beyond those covered in this introductory text.

11.3.1 Troughs and Ridges

In the previous section, we used WV imagery to identify dry and moist regions within the upper troposphere. Because WV imagery largely diagnoses upper-tropospheric moisture, it can also be used as a useful tracer of the upper-atmospheric wind and geopotential height pattern. Fig. 11.9 shows a full disk GOES-16 WV image (Fig. 11.9A) and a 250-hPa wind chart (Fig. 11.9B) at the same time. Focus on the region marked "1," located east of Lake Ontario. On the 250-hPa wind chart (Fig. 11.9B), there is a strong upper-tropospheric trough roughly centered over this location. Now examine the same area on the WV image (Fig. 11.9A); if we look closely, we can see that there is a cyclonic "dip" in the WV pattern. That is, the upper-tropospheric moisture follows the shape of the corresponding upper-tropospheric trough. Area "2," just off the west coast of South

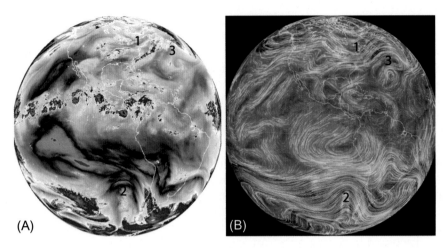

Fig. 11.9 At 18Z 3 July 2017: (A) GOES-16 full disk WV satellite imagery and (B) winds (*shaded*, streamlines) at 250 hPa. Corresponding areas 1, 2, and 3 are marked in *black* on each panel for the discussion of feature identification in this section. (A) Reproduced with permission from College of DuPage NEXLAB: GOES-16 imagery, available online at: http://weather.cod.edu/satrad/exper/ and (B) From EarthWindMap: 250-hPa winds, available online at: https://earth.nullschool.net/.

America (Fig. 11.9B), is an even stronger upper-tropospheric trough (air rotates clockwise around a trough in the Southern Hemisphere). When we examine the corresponding region on the WV image (Fig. 11.9A), we see spot the equatorward arc of WV as it moves cyclonically (clockwise) around the trough.

Finally, examine the 250-hPa ridge over the North Atlantic Ocean, marked with a "3" (Fig. 11.9B). WV arcs poleward as it travels anticyclonically around an upper-tropospheric ridge. If we focus on the nearby blue areas in Fig. 11.9A, we can see this anticyclonic arcing of moisture around the ridge.

Identifying troughs and ridges from a single WV image is not an easy task; generally, the analyst or forecaster should use WV animations to more simply locate this features. Regardless, WV imagery can be extremely useful in diagnosing upper-tropospheric patterns and as a source of comparison to a model analysis or forecast. For example, imagine you are making a 24- to 48-h forecast for a given region based on 00Z model data. One of the best ways to evaluate whether the model is diagnosing the current conditions correctly is to compare its upper-tropospheric analysis to a WV image. If the model is not accurately representing the current pattern, it is very unlikely it will be able to accurately replicate any future patterns.

11.3.2   Fog and Low Clouds

In Section 11.2, we briefly examined fog in Washington and Oregon and its respective signatures on both VIS and IR imagery. Recall that fog and low stratus are typically relatively thick, but located at or very close to the surface. The best procedures for employing satellite imagery to detect fog are to

- Use animations. Fog, particularly radiation fog, will typically not move, especially in valleys.
- During the day, use a combination of VIS and IR imagery. Fog and low stratus will appear relatively bright (thick) on VIS and dull gray (warm) on IR.
- Use surface (METAR) observations in conjunction with satellite imagery. Surface observations can be especially useful at night, when VIS imagery is not available, because IR imagery struggles to discern fog from the ground. Surface observations can also help to differentiate between fog, located at the ground, and stratus located just above the ground. These differences can be very important for visibility, especially to transportation sectors such as aviation.

In addition to the aforementioned radiation fog over Washington and Oregon, there appear to be scattered clouds off of the Oregon coast. We can tell these clouds are likely low stratus, because they are relatively thick (Fig. 11.6C) and warm (Fig. 11.7C), but not quite as warm as (slightly brighter on IR than) the fog over the land. If you were to animate these images, the stratus over the ocean would likely move; this is a common feature on the Pacific Coast, commonly called the "marine layer." The marine layer is sometimes referred to as advection fog or stratus, and unlike radiation fog, tends to move with the prevailing low-level wind.

11.3.3   Thunderstorms

Thunderstorms are a persistent hazard to life, property, and transportation, particularly aviation. Satellite imagery enables meteorologists to observe the development, evolution, and motion of thunderstorms in real time over much of the globe. For the novice forecaster, the best cues to look for when diagnosing thunderstorms are

- Very thick (bright white) clouds on VIS imagery.
- High (cold) cloud tops on IR imagery.
- Large amounts of moisture on WV imagery.

Additional examples of thunderstorms are provided in the practices exercises at the end of this chapter.

11.4 Questions and Practice Exercises

1. Briefly explain the differences in orbit between polar-orbiting and geostationary weather satellites.
2. How frequently will a polar-orbiting satellite pass over Toronto, Canada?
3. Name two limitations of geostationary satellites and discuss strategies you can use to mitigate these limitations.
4. It is 0630Z in January and you are trying to use satellite imagery to see if fog is forming over Charlotte, North Carolina. Discuss the type(s) of satellite imagery you would use and the best strategies to investigate fog formation.
5. On VIS satellite imagery, why are animations useful to identify radiation fog and snow cover?
6. For each of the following weather phenomenon, which type of satellite imagery should you use to best identify it? Explain your answers:
   * An upper-tropospheric cutoff cyclone.
   * Fresh snow cover.
   * A supercell thunderstorm.
   * Advection fog.
7. Use *both* the VIS and IR satellite images in Fig. 11.10 to identify the most likely type of clouds at each red numbered region on the VIS image.

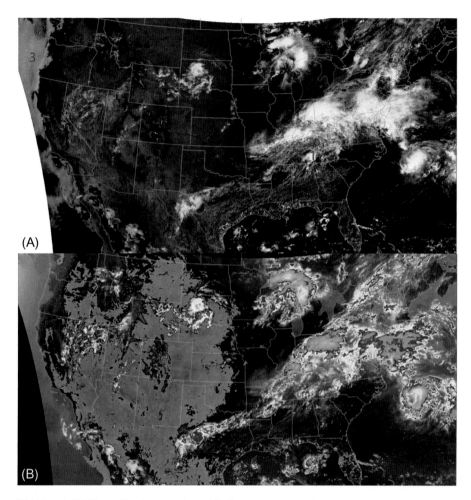

Fig. 11.10 (A) VIS and (B) IR satellite imagery from GOES-16 for use with Question 7. The numbered regions labeled in *red* on the VIS image represent the areas of interest in Question 7. Reproduced with permission from College of DuPage NEXLAB: GOES-16 imagery, available online at: http://weather.cod.edu/satrad/exper/.

12

RADAR IMAGERY

12.1 Introduction to Weather Radar

12.1.1 How Radar Works

Radar is an acronym that stands for *Ra*dio *D*etection *a*nd *R*anging. It was first developed for military use just before and during World War II. Today, radar is used for many different purposes such as aviation, defense, law enforcement, baseball pitch speed, and of course, weather. Unlike a satellite, a weather radar uses an active remote sensor; it emits a pulse of infrared radiation and then "listens" for the amount of energy that is reflected back toward it. The amount of power, or reflectivity, that is returned to the radar allows it to determine both the size and number of object(s) the radar beam hits.

To discuss the utility of radar in weather analysis and forecasting, we will use National Weather Service (NWS) Weather Surveillance Radar (WSR-88D, sometimes referred to as NEXRAD) imagery. The WSR-88Ds (Fig. 12.1A) are a network of radars in the United States and US Territories that emit a 10-cm wavelength beam. Other countries may use slightly different wavelengths, although the vast majority of operational radars operate in the 3–10-cm range. This means that weather radars can detect objects of precipitation size or greater, but not smaller objects such as cloud droplets.

Fig. 12.1B is a schematic of how a radar works. The radar dish, or transmitter, progressively emits a beam at a series of elevation angles; however, the radar never points either perfectly parallel to the ground (horizontally) or straight up (vertically). The transmitter also rotates in a 360° swath to complete scans in every direction. When the radar pulse hits a target (e.g., precipitation), a portion of the beam's energy will be scattered by that target, and another fraction of energy will be reflected back toward the transmitter. The more targets the beam hits, the more energy that is returned to the transmitter; this is

Synoptic Analysis and Forecasting. https://doi.org/10.1016/B978-0-12-809247-7.00012-0

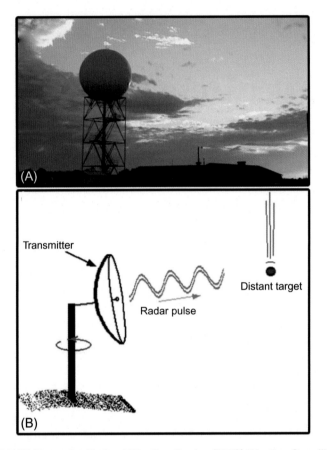

Fig. 12.1 (A) Photo of a National Weather Service (NWS) Weather Surveillance Radar (WSR-88D) and (B) schematic of how a weather radar works. (A) From the National Oceanic and Atmospheric Administration (NOAA): NOAA News, available online at: http://www.noaanews.noaa.gov/stories2013/images/WSR-88D_Tower.jpg and (B) Reproduced with permission from University of Illinois Weather World 2010: Radar Meteorology. Available online at: http://ww2010.atmos.uiuc.edu/guides/rs/rad/basics/gifs/dpmod1.gif.

why heavy precipitation appears as a stronger reflectivity echo than does light precipitation.

12.1.2 Doppler Radar

Although radar has been used for meteorological purposes since World War II, radar capabilities have advanced throughout the years. The original weather radars were primarily only able to detect location and intensity of precipitation. In the mid-late 1980s, US radars were upgraded to WSR-88Ds, which introduced Doppler principles to weather detection. The Doppler effect

(Doppler shift) is named after 19th-century Austrian physicist Christian Doppler and is based on the principle that the frequency (wavelength) of an electromagnetic pulse (wave) will change for an observer moving relative to the wave source. In meteorological terms, because the radar beam's targets (i.e., precipitation) are moving relative to the location of the radar, a Doppler radar is able to detect changes in frequency that allow us to detect whether the targets are moving toward or away from the radar.

The advent of Doppler radar allowed meteorologists to be able to determine wind velocity information, in addition to precipitation intensity and motion. This completely revolutionized our ability to detect and warn for life-threatening severe weather phenomena such as tornadoes and microbursts. In fact, before the late 1980s, microbursts (rapid downdrafts <4 km in width) were the cause of numerous fatal jet aircraft accidents. Doppler radar virtually eliminated this problem; in fact, not a single commercial airliner has crashed in the United States due to a microburst since Doppler radar became operational. As we will see in Section 12.2, Doppler radar has also vastly increased tornado warning lead times, because it allows meteorologists to see rotation within a thunderstorm. These days, most if not all operational weather radars across the globe employ Doppler principles, and the technology continues to improve over time.

12.1.3   Dual-Polarization Radar

One of the most recent advances in radar has been the deployment of Dual-Polarization ("Dual-Pol") technology. In the United States, all WSR-88D radars were upgraded to include Dual-Pol in 2013. Before Dual-Pol, radars sent out a single horizontal pulse; this pulse could measure the intensity and motion of precipitation, but struggled to discern precipitation type and nonmeteorological objects. The Dual-Pol upgrade allows WSR-88Ds to emit two simultaneous pulses, one horizontal and one vertical (Fig. 12.2A). Emitting two pulses enables the radar to obtain three-dimensional characteristics of its targets. This has resulted in considerable improvements for weather analysis and forecasting, including more accurate estimates of tornadoes, precipitation type, and precipitation accumulation.

We will explore how Dual-Pol radar variables can be used to diagnose precipitation type and tornadoes in Section 12.3. Here, Fig. 12.2B shows an example of radar-estimated precipitation accumulation from a rainfall event in Oklahoma. Although radar-estimated precipitation amounts should not be used as a substitute for rain gauge data, they allow meteorologists to quickly

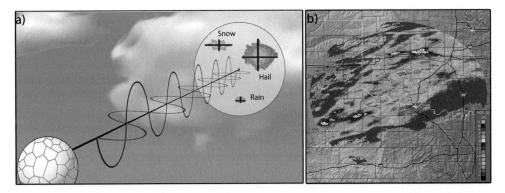

Fig. 12.2 (A) Schematic showing how Dual-Polarization (Dual-Pol) radar works, with both the horizontal and vertical pulses pictured and (B) example of Dual-Pol radar-estimated storm-total accumulated precipitation (inches, *shaded*). From (A) National Weather Service (NWS): Dual-Polarization radar, available online at: https://www.weather.gov/news/130425-dualpol/ and (B) NWS: Jetstream, an Online School for Weather, available online at: http://www.srh.noaa.gov/jetstream/doppler/precip.html.

get an idea of how much precipitation has fallen. This can be an important tool in thunderstorm flash flood situations, during which tens of millimeters of rain can fall in less than an hour, often between METAR stations and/or hourly reports.

12.1.4 US Radar Network

As a radar beam travels farther away from the transmitter, it is increasingly scattered by the atmosphere. Therefore radars have a limited horizontal and vertical range. In the horizontal, 10-cm radars can scan objects approximately 370 km away from the transmitter, although accuracy decreases with increasing distance due to scattering. As a result, most countries employ multiple overlapping radars, such that radar coverage is relatively uniform throughout. Fig. 12.3A shows the current WSR-88D network, not including Alaska, Hawaii, and overseas territories. Most of the 159 WSR-88Ds are owned and operated by NOAA, although others are maintained by the Department of Defense (DoD) or the Federal Aviation Administration (FAA). East of the Rockies, there is nearly 100% coverage, with considerable overlap of multiple radars in most locations. As examples, in Florida, the Melbourne (KMLB) and Tampa Bay (KTBW) radars overlap, while in Texas, the Corpus Christi (KCRP) and Brownsville (KBRO) radars overlap substantially. The overlap is deliberately designed so that if one radar requires maintenance, coverage can still be provided by a neighboring radar. In the Mountain West and parts of the West Coast, there are substantial gaps in radar coverage, such as in rural

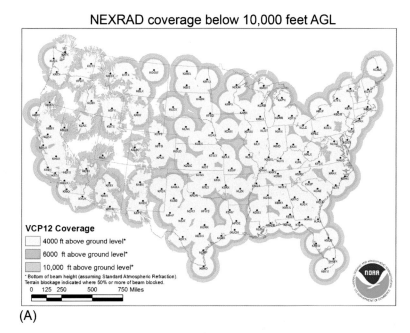

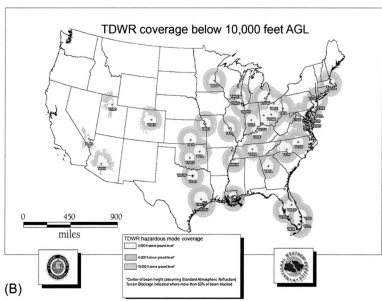

Fig. 12.3 Coverage maps of the United States (A) WSR-88D (NEXRAD) radar and (B) Terminal Doppler Weather Radar (TDWR) networks. From NWS: Radar Operations Center, available online at: https://www.roc.noaa.gov/WSR88D/Maps.aspx.

central Nevada and the sparsely populated Four Corners region. In the United States, most coverage gaps are limited to very unpopulated rural areas.

Recall how the advent of Doppler radar essentially eliminated the threat of microbursts to aviation; to enhance meteorologists' ability to detect microbursts and other thunderstorm-related phenomena in real time, the FAA has 45 Terminal Doppler Weather Radars (TDWRs) located at many of the largest airports in the United States and Puerto Rico (Fig. 12.3B). TDWRs have a smaller horizontal range than WSR-88Ds, but offer finer spatial resolution, which is useful for detecting small-scale weather features such as microbursts. TDWRs are especially useful at high-traffic airports with a large thunderstorm frequency, such as those in Florida, Texas, and throughout the Midwest and Northeast. Notice in Fig. 12.3B that west of the Rockies, only Salt Lake City, Las Vegas, and Phoenix have TDWRs; this is because thunderstorm frequency on the West Coast is the smallest anywhere in the US. Radar coverage in California, however, is still among the best in the United States, with 11 WSR-88Ds located throughout the state (Fig. 12.3A).

Not all countries have the widespread radar coverage that the United States enjoys. Canada, the second largest country in the world in terms of area, where approximately 70% of the population lives within 160 km (100 miles) of the US border, has a limited radar network outside of the major population centers. This can be challenging for forecasters trying to issue severe weather watches and warnings, particularly in polar regions where geostationary satellite data is not available. In regions where geostationary satellites can scan, the newest generation of satellites is helping to fill radar gaps with improved spatial and temporal resolution, and lightning mappers.

12.2 Common Types of Radar Imagery

12.2.1 Reflectivity

The unit of radar reflectivity is the decibel (dBz), which is a logarithmic measure of reflected energy. A logarithmic scale is used because of the large difference between transmitted and returned energy. If we did not use a logarithmic scale, we would require a color bar for values from 0.001 (light fog) to 36,000,000 (large hail)! On the dBz scale, the range of reflectivity values is much more manageable, from a minimum of −30 dBz for light fog to 75 dBz for large hail.

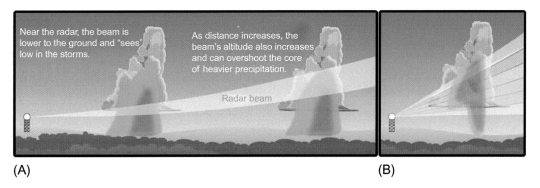

(A)   (B)

Fig. 12.4 Schematics of (A) base and (B) composite reflectivity radar scans. From NWS: Jetstream, an Online School for Weather, available online at: http://www.srh.noaa.gov/jetstream/doppler/refl.html.

In this chapter, we will discuss two types of reflectivity: base and composite. Fig. 12.4 illustrates how each type is derived. Recall that the lowest radar beam is still emitted at an angle; this angle is typically 0.5 degrees above the horizon (Fig. 12.4A), and the resulting scan is used as the "base" data. As the base beam travels farther away from the radar transmitter, it scans at a progressively higher altitude; this is why even base reflectivity will sometimes show virga (precipitation that evaporates before hitting the ground) far away from the radar.

Although a radar never points straight up,[1] it will scan at a series of progressively higher angles (Fig. 12.4B). Using an algorithm, computers can derive the largest reflectivity value within an entire vertical column at a given point, which is called composite reflectivity. Although composite reflectivity is not always representative of the precipitation closest to the ground, it does highlight the maximum reflectivity within a cloud, which is quite useful to meteorologists, particularly in the aviation sector. For example, the composite reflectivity may show 70 dBz, indicative of large hail; while this hail may mostly melt before hitting the ground, it is not something an aircraft should be flying anywhere near.

Fig. 12.5 shows two comparisons of base and composite reflectivity. Fig. 12.5A and B is radar images from a moderate to heavy precipitation event in New England; notice that while the two panels generally depict similar areal coverage, the reflectivity values are larger in the composite image (Fig. 12.5B) than in the base image (Fig. 12.5A). This is particularly evident over central

---

[1] The area directly above the radar (generally within 8 km) is called the "cone of silence," because the radar is not able to scan any objects there.

Base reflectivity   Composite reflectivity

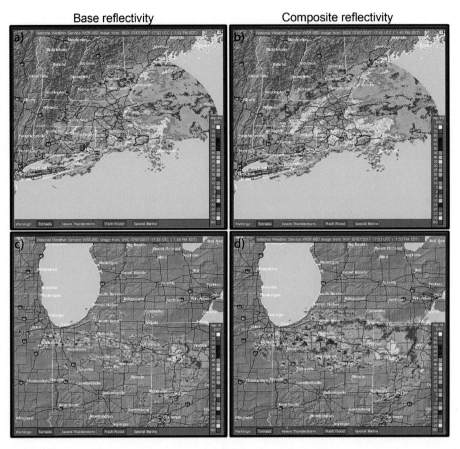

Fig. 12.5 NWS (left) base and (right) composite reflectivity (dBz, *shaded*) images on 7 July 2017, from the (A,B) Boston, Massachusetts WSR-88D and (C,D) Northern Indiana WSR-88D. Data from NWS: National Doppler Radar Sites, available online at: http://radar.weather.gov.

Massachusetts, where base reflectivity values are around 30–35 dBz, while composite reflectivity values are near 40 dBz. Meanwhile, the largest values of both base and composite reflectivity are colocated with the flash flood warning (green box) in southeastern Massachusetts.

Fig. 12.5C and D shows base and composite reflectivity for a thunderstorm outbreak over northern Indiana. The differences between base and composite reflectivity values are stark in these images, with base values in the strongest storms approximately 45–50 dBz while composite values are 55–60 dBz. The composite reflectivity over northern Indiana suggests at least a small hail threat at higher altitudes, while the hail threat on the ground is less, as evidenced by the general lack of severe thunderstorm warnings (only one yellow box in Fig. 12.5C, approximately half-way between Indianapolis and Fort Wayne, Indiana).

12.2.2 Radial Velocity

With the development of Doppler radar came the ability to detect the motion of objects scanned by radar. However, the only type of motion the radar can see is the portion of the wind that is moving directly *toward* or *away from* the radar itself. This component of the wind is called *radial velocity*, and is depicted in the schematic in Fig. 12.6A, in which the black arrows indicate a uniform southerly wind environment around the radar. As the radar beam scans in various directions (yellow lines), it will see only the component of the wind that is toward or away from the radar along each scan. For example, the radial velocity (blue arrow) toward the radar when the beam is pointed due southward (position 6) is larger (faster) than the radial velocity toward the radar when the beam is scanning to the southwest (position 7).

Computer algorithms will separate Doppler radial velocity (hereafter, simply velocity) into two different color scales, one for velocities moving *toward* the radar (green and/or blue), and one for velocities moving *away from* the radar (red). In Fig. 12.6A, because the actual wind is southerly, objects located to the north of the radar will be moving away from the radar,

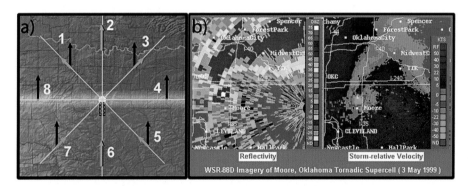

Fig. 12.6 (A) Schematic of how a Doppler Radar derives radial velocity and (B) WSR-88D base reflectivity (dBz, left) and storm-relative radial velocity (kt, right) during the 3 May 1999 Moore, Oklahoma EF-5 tornado. In (A), *yellow* indicates the radar beam emanating in several directions, *black arrows* indicate the actual wind, *blue arrows* show the radial velocity; *the red shaded area* indicates radial velocities oriented *away from* the radar, while the *green shaded area* indicates radial velocities oriented *toward* the radar. (A) From NWS: Jetstream, an Online School for Weather, available online at: http://www.srh.noaa.gov/jetstream/doppler/vel.html and (B) From NWS Storm Prediction Center (SPC), 2017: Frequently Asked Questions, available online at: http://www.spc.noaa.gov/faq/tornado/doppler.htm.

and thus are shaded red, while objects located south of the radar will move toward the radar and are shaded green. The analyst or forecaster must know where the radar is located relative to the objects it is scanning, in order to determine where objects are moving.

As is the case with radar reflectivity, there are two types of velocity products: base and storm relative. The difference between the two is that storm-relative velocity subtracts out the motion of the objects being scanned (e.g., a thunderstorm), so that we are, in effect, viewing the wind's motion as if the storm is stationary. This can be extremely useful for fast-moving phenomena such as squall lines or Mesoscale Convective Systems. A good rule of thumb is to use storm-relative velocity when storm motion is >30 kt, and base velocity when storm motion is <30 kt.

Fig. 12.6B shows base reflectivity and storm-relative velocity for the 3 May 1999 EF-5 Moore, Oklahoma tornado from the Norman, Oklahoma radar located due east of Moore. The reflectivity shows the classic hook-echo supercell shape just north and northwest of Moore. On the storm-relative velocity image, there is an area of red and green pixels north of Moore. The red pixels indicate motion away from the radar (in this case, toward the west), while the green pixels indicate motion toward the radar (in this case, toward the east). If you draw a diagram of or imagine inserting a pinwheel into this area, you will see that the pinwheel will spin counterclockwise, which is cyclonic in the Northern Hemisphere. Therefore we can conclude that neighboring red and green velocity pixels (called a "velocity couplet") are suggestive of thunderstorm rotation. This is how computer algorithms and meteorologists are able to detect a mesocyclone, or rotating updraft, in a thunderstorm. Although only about 40% of mesocyclones result in tornadoes, Doppler radar velocity has allowed meteorologists to increase warning lead times for storms that may produce tornadoes. In the next section, we explore how Dual-Pol radar technology has helped to reduce false alarm rates in tornado warnings.

12.2.3 Correlation Coefficient

Earlier, we learned how Dual-Pol technology enables the radar to send out two simultaneous pulses (one horizontal and one vertical), in order to retrieve a three-dimensional profile of objects scanned by the radar. Although Dual-Pol technology has provided meteorologists with several new products, we will only be

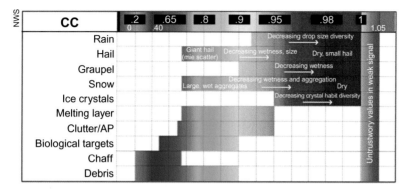

Fig. 12.7 Values of correlation coefficient (CC) and associated phenomena. From Schlatter, T., 2012. Weather queries. *Weatherwise*, **65(2)**.

discussing one here; correlation coefficient (CC) is a measure of how similarly the horizontal and vertical pulses are performing within a given pulse volume. To understand this concept, consider the three-dimensional shapes of both precipitation particles and nonmeteorological objects. Raindrops, and even small hail, are relatively spherical and uniform in shape, such that the horizontal and vertical radar pulses will behave similarly when they strike them. Objects that have a perfectly uniform shape have a CC value of 1, by designation. CC values <1 indicate objects that are progressively less uniform.

Fig. 12.7 shows the typical table of values and color shading used for CC. Notice that snowflakes, particularly wet snowflakes, and larger hail, have smaller CC values than raindrops or small hail. Snowflakes and larger hailstones tend to have unique and less uniform shapes; for example, the old adage "no two snowflakes are alike" is meteorologically factual. These properties have enabled meteorologists to use CC to discern precipitation type; as we will see in Section 12.3, CC can be incredibly useful in mixed and/or frozen precipitation events such as those producing ice pellets and freezing rain.

On the scale in Fig. 12.7, notice that nonmeteorological objects such as clutter, chaff, biological targets (e.g., birds, insects), and debris have the smallest CC values. Before Dual-Pol, it was difficult to differentiate precipitation from nonmeteorological objects on radar. However, CC, among other products, has made it a relatively simple exercise for trained meteorologists. In the next section, we will see how CC is used to identify tornadic debris and thereby reduce tornado warning false alarm rates.

12.3   Feature Identification

Weather radar has many applications beyond the few simple examples in this text. The reader is strongly encouraged to seek tutorials and learning modules on the latest technology, especially Dual-Pol radar. To provide a sense of how useful radar can be to meteorologists, we will now examine two important phenomena that that greatly affect life and property: tornadoes and frozen precipitation.

12.3.1   Tornadoes

Fig. 12.8 shows base reflectivity, base velocity, CC, and the damage path of the 2013 Moore, Oklahoma EF-5 tornado. A reflectivity hook echo (Fig. 12.8A) is clearly evident, while the velocity "couplet" is extremely strong (Fig. 12.8B). Now look at the circled area of blue shading on the CC image (Fig. 12.8C), using CC scale in Fig. 12.7; such a large blue area is almost always evidence of nonmeteorological objects. When low CC is colocated with a strong velocity couplet and an anomalously strong reflectivity

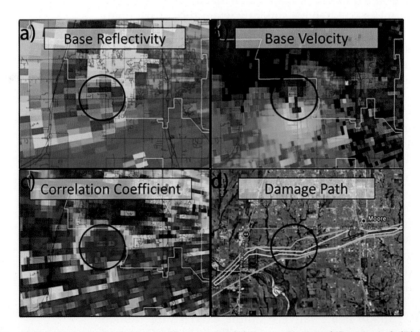

Fig. 12.8 For the 20 May 2013 Moore, Oklahoma EF-5 tornado: (A) base reflectivity (dBz, *shaded*), (B) base radial velocity (kt, *shaded*), (C) CC (*shaded*), and (D) damage path. The color scales are similar to those used in Fig. 12.6 for reflectivity and radial velocity, and as in Fig. 12.7 for CC. From the NWS Warning Decision Training Division: Dual-Polarization radar training for NWS partners, available online at: http://www.wdtb.noaa.gov/courses/dualpol/Outreach/.

signature (often called a "debris ball"), it is indicative of tornado debris. When used with velocity, CC is so accurate at identifying debris that meteorologists defined it as a Tornadic Debris Signature (TDS). While not all tornadoes lift debris (consider a tornado in an open field for example), the TDS has greatly increased the identification and warning rate for tornadoes that do. In the past, meteorologists could only say there was a rotating thunderstorm, based on the velocity couplet. Now, with the advent of the TDS, warning lead times have increased and false alarm rates have dropped.

12.3.2 Frozen Precipitation

In the midlatitudes and polar regions, frozen precipitation can cause major disruptions in transportation and commerce, as well as damage to life and property. Forecasting mixed or frozen precipitation (e.g., ice pellets, freezing rain) is one of the most difficult tasks for a forecaster, because of all of the competing warming and cooling processes. In the past, observations of frozen precipitation were mostly limited to METAR reports, twice-a-day radiosonde measurements, and aircraft reports. Dual-Pol radar products, including but not limited to CC, have revolutionized how quickly meteorologists can identify various types of frozen precipitation. For the sake of brevity, we only discuss a CC example here, but the reader is encouraged to learn about the utility of Differential Reflectivity (ZDR) and other useful Dual-Pol radar products.

Fig. 12.9 shows base reflectivity and CC from a cold-season precipitation event in the New York City metro area. Using only reflectivity (Fig. 12.9A), we can discern heavier areas of snow by looking for dark green, and possible areas of mixed precipitation (melting) where there is yellow, a phenomenon called "radar bright banding."[2] In the past, meteorologists would need to correlate bright banding areas with surface observations to confirm any potential frozen and/or mixed precipitation; with Dual-Pol, an extra data source is not necessary. The CC image in Fig. 12.9B clearly shows CC values near 1 throughout much of the snow region in Connecticut, which is typical for relatively dry snow. However, over New York City and south of Long Island, CC values are <1 (Fig. 12.7), suggesting nonuniform shapes and melting/

---

[2] Radar bright banding is a term used to describe how ice particles, especially ice pellets and melting snowflakes, often appear on reflectivity. Because ice particles are more reflective than snow and sometimes even rain, it can seem that precipitation is heavier (larger reflectivity values) when it is actually just mixed or frozen precipitation.

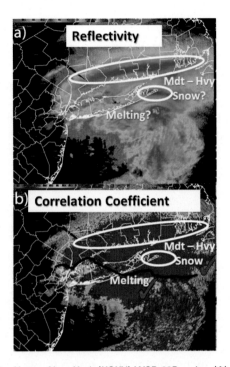

Fig. 12.9 From the Upton, New York (KOKX) WSR-88D radar: (A) base reflectivity (dBz, *shaded*) and (B) CC (*shaded*), for a frozen precipitation event. The color scales are as in Fig. 12.6 for reflectivity and as in Fig. 12.7 for CC. Modified from NWS: Life-saving technology upgrade coming to a Doppler radar near you, available online at: https://www.weather.gov/news/080212-dual.

mixed precipitation. As with the TDS, Dual-Pol has increased warning lead times and forecaster confidence during ice pellet and freezing rain situations. Ground truth (i.e., surface observations) is always useful to a forecaster, but radar also serves an important role in advancing predictability and enhancing public safety.

12.4 Questions and Practice Exercises

1. Why is a logarithmic scale used to display radar reflectivity and what is the unit of radar reflectivity called?
2. Explain the key difference in remote sensing techniques between a weather satellite and weather radar.
3. Explain the difference between base and composite reflectivity. Why would a forecaster want to preferentially use composite reflectivity on a particular thunderstorm?

4. The hazard to aviation due to which weather phenomenon has been practically eliminated by the advent of Doppler radar?

5. Explain the difference between base velocity and storm-relative velocity. What are the proper situations in which to use each one?

6. How does Dual-Pol technology change how a weather radar operates?

7. Fig. 12.10 shows base reflectivity, base velocity, and CC for a past tornado in Kentucky. Where on the map is the tornado located? Explain how you know.

8. Fig. 12.11 shows base reflectivity and CC for a frozen precipitation event in New England. Determine the areas of (a) snow and (b) melting/mixed precipitation. Explain your answer.

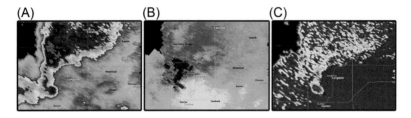

Fig. 12.10 From the Louisville, Kentucky (KLVX) WSR-88D radar: (A) base reflectivity (dBz, *shaded*), (B) base velocity (kt, *shaded*) and (C) CC (*shaded*), for a tornado event. The color scales are similar to those used in Fig. 12.6 for reflectivity and radial velocity, and as in Fig. 12.7 for CC. Modified from NWS Louisville: NWS Doppler radar Dual-Pol—Tornado debris, available online at: https://www.weather.gov/lmk/nws_radar_dualpol_tordebris.

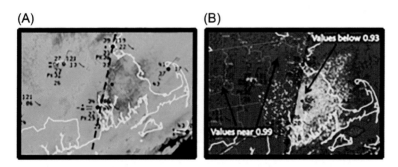

Fig. 12.11 From the Boston, Massachusetts (KBOX) WSR-88D radar: (A) base reflectivity (dBz, *shaded*) and (B) CC (*shaded*), for a frozen precipitation event. The color scales are as in Fig. 12.6 for reflectivity and as in Fig. 12.7 for CC. Modified from NWS: NWS Eastern Region completes Dual-Polarization upgrade, available online at: https://www.weather.gov/news/121712-eastern.

13

THERMODYNAMIC DIAGRAM BASICS

13.1 Skew-T Log-P Diagram

13.1.1 Introduction

Thermodynamic diagrams are an incredibly useful tool for weather analysis and forecasting. By the conclusion of the next two chapters, the reader should be able to diagnose most weather situations simply by examining a thermodynamic diagram. Although there are a few types of thermodynamic diagrams (e.g., tephigram) used across the world, here we will focus on the Skew-T Log-P diagram. The Skew-T Log-P diagram allows the analyst or forecaster to evaluate:

- How atmospheric variables (i.e., temperature, moisture) change with height.
- The vertical profile and stability of the atmosphere, particularly the troposphere, where most weather occurs.
- Relatively complex calculations of thunderstorm indices and other parameters, by using the multiple sets of reference lines on the chart.

In this section, we will introduce each set of lines on the Skew-T Log-P diagram and explain what they physically represent. In Section 13.2, we will learn how to use a thermodynamic diagram to lift an air parcel, evaluate atmospheric stability, and compute several indices that can diagnose thunderstorm probability and potential severity.

13.1.2 Components

The first set of lines are horizontal pressure lines (isobars), as seen in Fig. 13.1A. The isobars are irregularly spaced because pressure is plotted on a logarithmic scale, thus "Log-P." The distance between each 100 hPa of pressure becomes larger with

Synoptic Analysis and Forecasting. https://doi.org/10.1016/B978-0-12-809247-7.00013-2

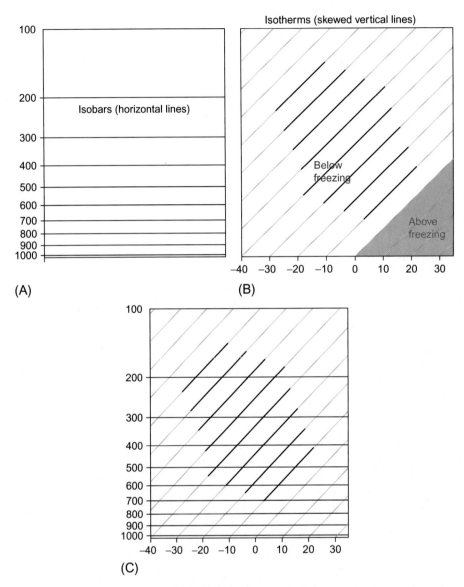

Fig. 13.1 Pressure and temperature lines on a Skew-T Log-P diagram: (A) horizontal isobars (hPa), (B) skewed isotherms (°C), with above- and below-freezing regions identified, and (C) isobars and isotherms plotted together.

increasing height (Fig. 13.1A) because air density decreases with increasing altitude, meaning a given volume of pressure (e.g., 100 mb) will be more expansive (taller) as density decreases.

Fig. 13.1B shows lines of constant temperature (isotherms) that skew across the diagram from bottom left to top right, thus "Skew-T." Temperature increases to the right (scale in

Fig. 13.1B), such that if we examine the 0°C isotherm in Fig. 13.1B, above-freezing temperatures are located to the right, while below-freezing temperatures are to the left. Finally, Fig. 13.1C shows the isobars and isotherms plotted together.

Fig. 13.2 details the remaining three sets of lines on the Skew-T Log-P diagram. In Fig. 13.2A, there is a set of dashed blue lines

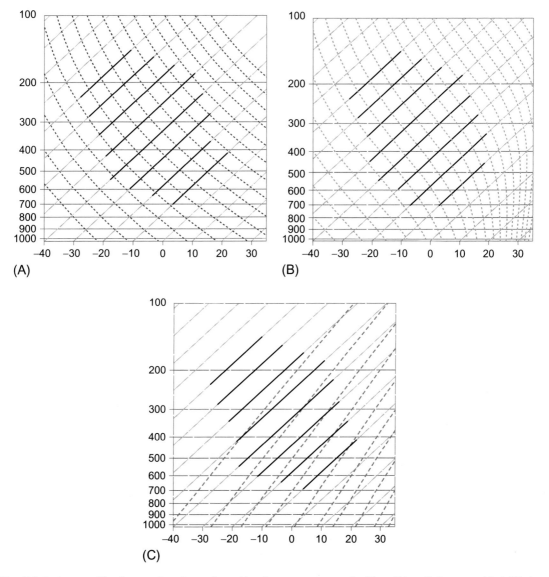

(A)

(B)

(C)

Fig. 13.2 Isobars and isotherms plotted together with other components of a Skew-T Log-P diagram added: (A) dry adiabats (*blue dashed*), (B) moist adiabats (*green dashed*), and (C) lines of constant mixing ratio (*brown dashed*).

drawn from lower right to upper left; these are called "dry adiabats" and they represent the dry adiabatic lapse rate (DALR). Recall that an adiabatic process is one during which the temperature of an air parcel can only cool or warm through expansion or compression, respectively. As air parcels rise, they expand and therefore cool adiabatically; as air parcels sink, they compress, and therefore warm adiabatically. The DALR is the rate at which *unsaturated* [relative humidity (RH) < 100%] air parcels cool or warm and is a constant $9.8°C \, km^{-1}$ at any temperature or pressure level. In Section 13.2, we will learn how to use the dry adiabats to lift an air parcel on a Skew-T Log-P diagram.

The curved green dashed lines in Fig. 13.2B are moist adiabats, representative of the moist adiabatic lapse rate (MALR). The MALR is the rate at which *saturated* (RH = 100%) air parcels cool or warm, but unlike the DALR, it is not constant. When rising saturated parcels cool adiabatically, they release latent heat, which offsets some of the adiabatic cooling. Therefore the MALR is smaller than the DALR and its magnitude is dependent on temperature, because warmer saturated parcels release more latent heat. At warmer temperatures, the MALR is less steep than it is at colder temperatures (Fig. 13.2B), indicating that as warmer saturated air parcels rise, they cool more slowly than colder saturated air parcels. In general, $6°C \, km^{-1}$ can be thought of as an average MALR value, but as Fig. 13.2B shows, the MALR is larger at colder temperatures and higher altitudes. In fact, if an air parcel is cold and/or high enough, the MALR will nearly equal the DALR.

Finally, moist adiabatic rising and sinking is often referred to as a *psuedoadiabatic* process, because it is not reversible. When a saturated parcel sinks, there is typically less cooling due to evaporation than there is latent heat due to condensation when that same air parcel rises. The result is that a saturated parcel can arrive back at its original pressure level with a different temperature and/or dew point than when it started rising.

The final set of lines are the dashed brown lines that skew up and to the right in Fig. 13.2B. These are called lines of constant mixing ratio, and can be identified by their slope, which is not as steep as the isotherms. Recall that mixing ratio is a measure of atmospheric humidity, defined as the ratio of the mass of water vapor to mass of dry air in an air parcel. The typical units are grams of water vapor per kilogram of dry air $(g \, kg^{-1})$. As evidenced by the lack of curvature in the lines, mixing ratio is conserved for a rising unsaturated air parcel. We further explore this point in Section 13.2, where we will use mixing ratio lines to help lift an air parcel to the level of condensation.

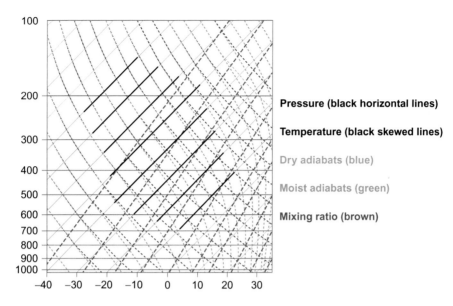

Fig. 13.3 A complete Skew-T Log-P diagram, incorporating all of the components in Figs. 13.1 and 13.2.

Fig. 13.3 shows the complete Skew-T Log-P diagram. Although the number of lines can seem daunting at first, once proper identification and procedures are learned, it becomes one of the most powerful tools in weather analysis and forecasting. For the majority of this chapter, we will use the color scheme shown in Fig. 13.3, but it is important to focus on the slopes of the lines, not the colors. Different websites and sources may use different colors for each set of lines than those in Fig. 13.3. However, if the analyst or forecaster can recognize, for example, that isobars are horizontal, and isotherms skew up and to the right, more steeply than the mixing ratio lines, he or she will always be able to correctly identify each set of contours.

13.2 Thermodynamic Variables and Indices

13.2.1 Lifting a Parcel

Recall that an air parcel is an imaginary small volume of air that does not exchange heat with the surrounding environment. As an air parcel rises to lower pressure, it expands and cools adiabatically; as an air parcel sinks, it compresses and warms adiabatically. The process of lifting an air parcel will demonstrate how a Skew-T Log-P diagram can be used to diagnose cloud base height, atmospheric stability, and thunderstorm potential. In this section,

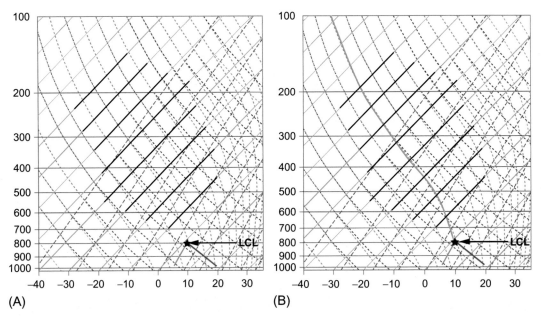

Fig. 13.4 Schematics showing how to lift an air parcel using a Skew-T Log-P diagram: (A) the air parcel is lifted (*solid red line*) from the surface temperature parallel to the nearest dry adiabat to the lifted condensation level (LCL, *black star*), where it meets the surface dew point lifted parallel (*solid green line*) to the nearest line of constant mixing ratio, and (B) the air parcel is lifted (*solid light blue line*) from the LCL, parallel to the nearest moist adiabat.

we will focus on how to lift an air parcel from the surface; however, it is worth remembering that certain weather conditions such as elevated convection may necessitate starting the lifting process at a higher altitude.

Fig. 13.4A shows a situation in which the surface pressure is 988 hPa, surface temperature is 18°C (bottom of red line), and surface dew point is 4°C (bottom of green line). The air parcel is unsaturated at the surface because the temperature and dew point are not equal (RH < 100%). In order to start lifting a parcel, the surface pressure, temperature, and dew point must be known beforehand, usually from radiosonde and/or surface observations. Because the air parcel is unsaturated, its temperature will cool at the DALR; on a Skew-T Log-P diagram, this is accomplished by starting at the surface temperature and paralleling the nearest dry adiabat (see red path in Fig. 13.4A). The dew point is lifted parallel to the lines of constant mixing ratio, and also cools with height (green path in Fig. 13.4A), but much more slowly (approximately 2°C km^{-1}), as the parcel's mixing ratio is conserved. Because the temperature cools at a much faster rate than the dew point, the two lines eventually intersect. At this pressure level, the *lifting condensation level* (LCL), temperature, and dew point are equal (RH = 100%) and the air parcel is saturated. The LCL is therefore the altitude where most cloud bases form, because cloud formation requires saturation and condensation. In Fig. 13.4A, the LCL and cloud bases are located at 800 hPa.

Once an air parcel is lifted to the LCL, it becomes saturated and will remain that way as it continues to rise. Saturated air parcels, of course, cool or warm at the MALR. In Fig. 13.4B, the blue air parcel path starting at the LCL parallels the nearest moist adiabat to the top of the diagram. As the air parcel gets to higher altitude (lower pressure), the moist adiabats become incrementally more parallel to the dry adiabats; this is indicative of progressively less condensation and latent heat release at colder temperatures and higher altitudes.

Now that we have learned how to lift an air parcel to and above the LCL (called the "air parcel path"), we will review several other important parameters on the Skew-T Log-P diagram that are particularly useful in thunderstorm forecasting. Figure 13.5 shows a new scenario in which the surface temperature and dew point are both warmer than in Fig. 13.4 and the LCL is located at approximately 900 hPa. In addition, there are two bold black lines in both panels of Fig. 13.5; the right- (left-) hand line is representative of the environmental temperature (dew point), as measured by a

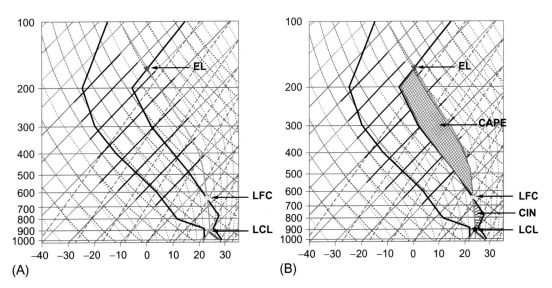

Fig. 13.5 (A) The air parcel is lifted (*solid light blue line*) from the LCL, parallel to the nearest moist adiabat with the level of free convection (LFC, *yellow star*), equilibrium level (EL, *pink star*) indicated, and (B) The air parcel is lifted (*solid light blue line*) from the LCL, parallel to the nearest moist adiabat with convective available potential energy (CAPE, *red hatched area*), convective inhibition (CIN, *purple hashed area*), and surface wet-bulb temperature (T_w, *bottom of orange line*) also indicated in the figure. In both panels, environmental temperature and dew point (°C) are indicated by the right- and left-hand *bold black lines*, respectively.

radiosonde. The change in environmental temperature with height is called the environmental lapse rate (ELR). Unlike the DALR and to some degree the MALR, the ELR is rarely near constant for more than a small vertical layer. However, vertical profiles of environmental temperature and moisture are extremely useful for analyzing specific weather conditions (see Chapter 14). For the remainder of this section, we will mostly use the ELR as a point of comparison between the environment and air parcel on the Skew-T Log-P diagram.

After the air parcel saturates at the LCL, it continues to rise and cool at the MALR. If and when the air parcel temperature becomes warmer than the environmental temperature, it has reached the *level of free convection* (LFC), which is approximately 630 hPa in Fig. 13.5A. Above the LFC, the parcel is warmer than the environment, meaning it will rise freely on its own without requiring a lifting mechanism (e.g., upper-level divergence, surface convergence). This is a key factor in thunderstorm formation; although lifting mechanisms are required to lift air parcels to the LFC, they typically cannot produce a 40,000 ft. cumulonimbus cloud. Much of the work in producing such a cloud must be done by the parcel rising on its own, because it is warmer than the surrounding environment. This situation is called *conditional instability*, in which a *saturated* parcel is unstable, because the magnitude of the ELR exceeds that of the MALR. Given an initial push upward, conditionally unstable saturated air parcels will continue to rise away from their initial height, and clouds will continue to grow taller.

Above the LFC, the air parcel rises freely until its temperature once again becomes cooler than the temperature of the environment, which occurs at the *equilibrium level* (EL). In Fig. 13.5A, the EL is located at approximately 180 hPa. The EL can be used as a rough estimate of a thunderstorm's top height. In Fig. 13.5A, the EL is found above the point (~200 hPa) where the stratospheric inversion in the ELR begins (Chapter 14); this indicates that the cumulonimbus has an overshooting top extending into the stratosphere.

13.2.2 Convective Indices and Surface Wet-Bulb Temperature

The area between the parcel path and ELR in the region from the LFC to the EL is called *convective available potential energy* (CAPE), hatched in red in Fig. 13.5B. CAPE is a measure of the amount of atmospheric instability and can be thought of as the amount of energy available for a thunderstorm. Because CAPE is an area, precise values are difficult to calculate by hand, but fortunately, modern computers are very adept at making quick CAPE calculations. The larger CAPE is, the stronger thunderstorms have the *potential* to be. However, it is important to remember that no amount of CAPE guarantees that thunderstorms will form; in addition to instability, they require adequate moisture and a lifting mechanism for the parcel to reach the LFC.

The purple hashed area located below the LFC in Fig. 13.5B is called *convective inhibition* (CIN). CIN represents the area on a sounding where the temperature of the air parcel is cooler than the temperature of the environment, indicative of a stable layer. To overcome CIN (sometimes called the "cap" or "capping inversion"), an air parcel must be pushed to the LFC by a lifting mechanism. More commonly, thunderstorms form after CIN has eroded, due to either surface heating, cooling aloft, or both.

We will now examine three additional variables and two additional thunderstorm indices that are useful for weather analysis and forecasting. In Fig. 13.5B, focus on the purple line extending from the LCL back down to the surface; this demonstrates how to calculate the surface *wet-bulb temperature (T_w)*. Recall that wet-bulb temperature is the coolest temperature that can be reached after all available water is evaporated into the air. In Chapter 15, we will see that T_w can be quite useful for temperature forecasting during heavy precipitation.

Fig. 13.6 shows the same environmental temperature and dew point sounding as in Fig. 13.5, but without the air parcel path. Also marked on Fig. 13.6 are the *convective condensation level* (CCL) and the *convective temperature* (CT). The CCL is similar to the LCL but is a more accurate measure of cloud base height in convective (thunderstorm) situations. To calculate the CCL, lift the air parcel from the surface dew point parallel to the nearest line of constant mixing ratio until the parcel path (solid green line in Fig. 13.6) intersects the temperature sounding (ELR, right-hand bold black line in Fig. 13.6). The CCL is also a pressure level, which in the case of Fig. 13.6 is approximately 890 hPa. In our example, the CCL and LCL (900 hPa) are within approximately 10 hPa of each other; it is generally rare for them to differ by >50 hPa. To find the CT, move from the CCL back down to the surface pressure, parallel to the nearest dry adiabat. Physically, the CT represents the temperature the surface needs to be heated to in order for convection to spontaneously initiate. However, convection can frequently initiate without the surface temperature warming to the CT, provided a lifting mechanism can overcome the CIN. CT is a useful forecasting tool in places such as Florida in summer, where surface heating is an important forcing mechanism for convection.

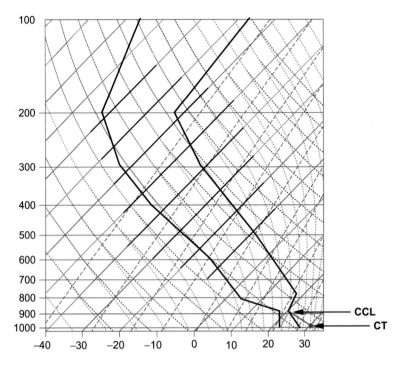

Fig. 13.6 Convective condensation level (CCL) and convective temperature (CT) are indicated in the figure. The CCL is found by lifting the parcel (*solid green line*) parallel to the nearest line of constant mixing ratio until it intersects the environmental temperature. The CT is found by descending (*solid red line*) dry adiabatically from the CCL back to the surface pressure. Environmental temperature and dew point (°C) are indicated by the right- and left-hand *bold black lines*, respectively.

In addition to CAPE, lifted index (LI) can be a useful measure of the amount of instability in the atmosphere. As shown in Fig. 13.7, LI is calculated by lifting an air parcel to 500 hPa, then subtracting the temperature of the parcel (T_P) from the environmental (sounding) temperature (T_E). If LI < 0, the parcel is warmer than the surrounding environment, indicating that the atmosphere is unstable, and the parcel will rise freely. Negative LI values and large CAPE values typically go together (Table 13.1); both are measures of *potential thunderstorm intensity*. Like CAPE, LI in no way guarantees that a thunderstorm will form; sufficient moisture and a lifting mechanism are still required. The primary limitation to LI is that it is only calculated at a single pressure level (500 hPa), while CAPE is a much more comprehensive measure of instability. Table 13.1 correlates several ranges of CAPE and LI values, and the associated degree of instability. Generally, a CAPE of >2000 and LI < −5 are

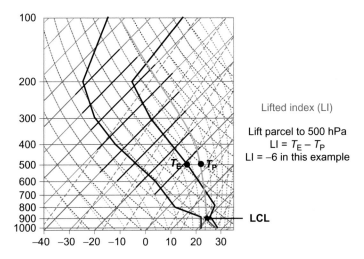

Fig. 13.7 The air parcel is lifted (*solid light blue line*) from the LCL, parallel to the nearest moist adiabat with the calculation procedure and result for lifted index (LI) shown in the figure. The environmental (T_E) and parcel (T_P) temperatures (°C) at 500 hPa are labeled on the diagram. Environmental temperature and dew point are indicated by the right- and left-hand *bold black lines,* respectively.

Table 13.1 Specific Value Ranges of CAPE and LI, and the Associated Degree of Instability

CAPE (J kg^{-1})	Lifted Index (LI)	Degree of Instability
500–1000	0 to −2	Marginally unstable
1000–2000	−3 to −5	Moderately unstable
2000–3500	−6 to −8	Very unstable
3500+	<−8	Extremely unstable

More instability indicates potential for stronger thunderstorms.

considered very unstable; if a thunderstorm forms under these atmospheric conditions, it is likely to be strong or severe. In extreme instability situations such as the severe thunderstorm season in the U.S. Great Plains and Canadian Prairies, CAPE can be as large as 5000–7000 J kg^{-1}, with corresponding LI values of −12 to −15.

The final index discussed in this chapter is the *K*-index (*K*), which is best used as a measure of the probability of thunderstorm occurrence. Fig. 13.8 shows the formula for *K*; each variable in the

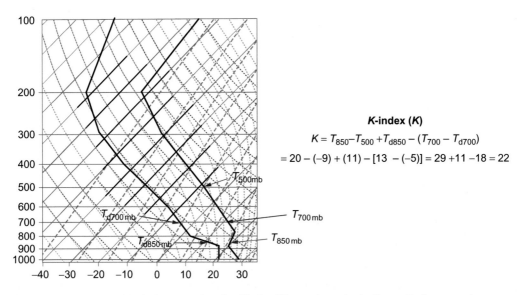

Fig. 13.8 The variables and calculation procedure for *K*-index (*K*) are shown in the figure. Environmental temperature and dew point (°C) are indicated by the right- and left-hand *bold black lines,* respectively.

formula is based on environmental temperature and/or dew point, and is not related to the lifted air parcel path. To calculate *K*, the ELR between 850 and 500 hPa (T_{850}–T_{500}) is first assessed, followed by adding the dew point at 850 hPa (T_{d850}) and subtracting the 700-hPa dew point depression at 700 hPa (T_{700}–T_{d700}). Although *K* does incorporate a measure of lower- and mid-tropospheric stability (ELR), it is primarily a way to assess the amount of lower-troposphere moisture, a necessary ingredient for thunderstorm formation. As a result, *K* is best used to discern the probability of thunderstorm occurrence, particularly ordinary ("air mass") thunderstorms in the warm season. *K* should not be used to assess potential thunderstorm severity, as CAPE and LI are much better indices for that purpose. Table 13.2 shows specific ranges of *K* and the associated thunderstorm probabilities. When forecasting, a good rule of thumb is to look for *K* > 30, which generally indicates that—given a lifting mechanism—thunderstorms are more likely than not to occur.

13.2.3 Complete Skew-T Log-P Analysis

Fig. 13.9 shows two radiosonde soundings from July 2017, from (A) Aberdeen, South Dakota and (B) Jacksonville, Florida. The color scheme in Fig. 13.9 is different than the Skew-T

Table 13.2 Specific Value Ranges of *K*, and the Associated Probability of Thunderstorm Occurrence

K-index (K)	Thunderstorm Probability of Occurrence (%)
<15	0
15–20	<20
21–25	20–40
26–30	40–60
31–35	60–80
36–40	80–90
>40	>90

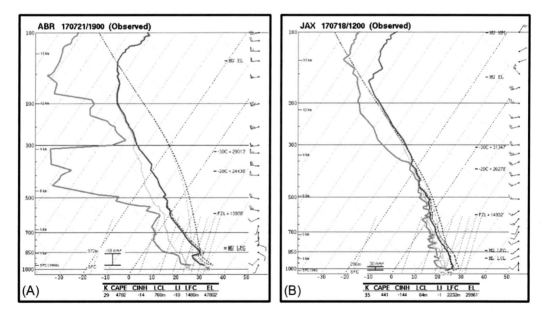

Fig. 13.9 Radiosonde soundings from July 2017 at (A) Aberdeen, South Dakota and (B) Jacksonville, Florida. In each panel, the environmental temperature (dew point) is plotted with a *solid red* (*green*) *line*, wind barbs (kt) are plotted on the right-hand side, and the approximate lifted parcel path is the *dark red dashed line*. Modified from the National Weather Service (NWS) Storm Prediction Center (SPC): Observed sounding archive, available online at http://www.spc.noaa.gov/exper/soundings/.

Log-P charts we have been using throughout this chapter; for example, the environmental temperature and dew point are plotted in red and green, respectively, instead of bold black. This is why it is crucial to commit the slopes of each set of lines to memory, so they can be readily identified regardless of the color scheme.

Examining the parcel paths (dark red dashed lines) and convective indices on each panel, we see that the Aberdeen sounding is much more unstable than the Jacksonville sounding. The CAPE at Aberdeen is 4702 J kg^{-1}, while at Jacksonville it is only 441 J kg^{-1}; correspondingly, the LI is much more negative (-10) at Aberdeen than at Jacksonville (-1). These values indicate that if thunderstorms form, they have the potential to be much stronger at Aberdeen than Jacksonville, and in fact, severe thunderstorms moved through eastern South Dakota shortly after the sounding was taken.

Unlike LI and CAPE, K is actually larger at Jacksonville (35) than at Aberdeen (29), indicative of the larger amount of moisture present in the lower troposphere at Jacksonville (Fig. 13.9B). $K > 30$ is relatively common in summer in Florida, where the lower troposphere is typically quite moist. For the analyst or forecaster, this discrepancy in convective indices can be best summed up as: the probability of thunderstorm occurrence is larger at Jacksonville, but if a thunderstorm occurs, it is much more likely to be strong or severe at Aberdeen. Finally, it is useful to compare the LCL heights between the two soundings. At Jacksonville, where the atmosphere is essentially saturated at the surface, the LCL height is only 64 m (~200 ft.) above the surface, while at Aberdeen it is >700 m (~2300 ft.) high. This is strongly suggestive of very low cloud bases at Jacksonville.

In Chapter 14, we will take the lessons learned here about Skew-T Log-P diagrams and apply them to many different types of soundings and weather scenarios. Before doing so, it is highly recommended that the novice analyst or forecaster practice plotting and interpreting thermodynamic diagrams. Familiarity and expertise comes with repetition, and sounding interpretation is one of the most powerful tools we have in meteorology.

13.3 Questions and Practice Exercises

1. Why is the MALR smaller than the DALR?
2. Of DALR, MALR, and ELR, which one is always a constant value? What is the value of this constant?

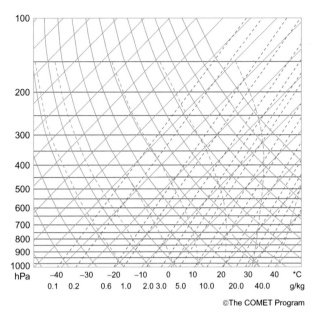

©The COMET Program

Fig. 13.10 Blank Skew-T Log-P diagram for use with Questions 3 and 4. On this diagram, the horizontal isobars are *dark blue*, the skewed isotherms are *solid pink*, the dry adiabats are *solid green*, the moist adiabats are *dashed green*, and the lines of constant mixing ratio are *dashed purple*. Courtesy of the COMET website at: http://meted.ucar.edu/ of the University Corporation for Atmospheric Research (UCAR), sponsored in part through cooperative agreement(s) with the National Oceanic and Atmospheric Administration (NOAA), US Department of Commerce (DOC). ©1997–2016 University Corporation for Atmospheric Research. All Rights Reserved.

3. Using the blank Skew-T Log-P diagram in Fig. 13.10 or one provided by your instructor, plot the environmental temperature and dew point data listed in Table 13.3.
4. For your plotted sounding in Question 3, identify or calculate each of the following parameters. Show your work.
   • LCL
   • LFC
   • EL
   • CCL
   • CT
   • CAPE
   • CIN
   • LI
   • K

Table 13.3 Radiosonde Measurements of Pressure (hPa), Temperature (°C), and Dew Point (°C), for Use With Questions 3 and 4

Pressure (hPa)	Temperature (°C)	Dew Point (°C)
966	27.4	22.4
937.6	24.1	19.6
925	22.8	18.6
905.6	20.9	17.7
850	16.6	14.6
785.4	17.3	0.5
700	9.6	−6.4
653.5	4.4	−9.7
586	−3.9	−14.9
500	−11.7	−27.7
400	−23.5	−39.5
300	−39.9	−58.9
250	−49.3	−70.3
200	−54.7	−71.7
150	−54.7	−72.7
100	−64.7	−78.7

5. What is the probability of occurrence of a thunderstorm if $K = 37$?
6. On some July day in Saskatoon, Saskatchewan, the LI is −7 and the CAPE is 3487 J kg^{-1}. How unstable is the atmosphere and what do these values indicate about potential thunderstorms?

THERMODYNAMIC DIAGRAM INTERPRETATION

14.1 Weather and Forecasting Diagnostics

Now that we have learned how to plot Skew-T Log-P diagrams, draw the air parcel path, and calculate convective indices, this chapter introduces sounding interpretation for many different weather situations. Soundings are an extremely powerful weather analysis and forecasting tool that can provide a great amount of information. By the conclusion of this chapter, the reader should be able to use a sounding at a given time and location to diagnose the state of the atmosphere without needing any supplemental analysis tools.

14.1.1 Cloud Layers

One of the simplest things to identify in an atmospheric sounding is a cloud layer. Recall that when the temperature and dew point are equal, relative humidity (RH) is 100%, which indicates saturation and condensation. A good rule of thumb is that saturation and condensation, and thus cloud formation, are likely when RH is ≥70%, which can be approximated by a dew point depression of ≤3°C.

Fig. 14.1 shows a sounding from 00Z 6 February 2010 during the infamous Mid-Atlantic "Snowmageddon" Blizzard.[1] From the surface to approximately 525 hPa, the atmosphere is completely saturated, i.e., the temperature and dew point are essentially equal. Therefore we can say with good certainty that

---

[1] A summary of the 2010 Mid-Atlantic Snowmageddon Blizzard is available from Weather Underground, online at: https://www.wunderground.com/blog/JeffMasters/top-us-weather-event-of-2010-snowmageddon.

Synoptic Analysis and Forecasting. https://doi.org/10.1016/B978-0-12-809247-7.00014-4

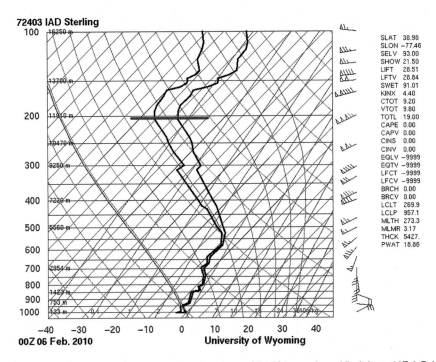

Fig. 14.1 Radiosonde sounding from Dulles International Airport (KIAD) in northern Virginia at 00Z 6 February 2010, during the Mid-Atlantic "Snowmageddon" Blizzard. The *thick black line* on the right (left) side represents environmental temperature (dew point) and the *thick red line* represents the approximate level of the tropopause. Wind is plotted on the far right in kt. Reproduced with permission from the University of Wyoming Department of Atmospheric Science sounding archive, available online at: http://weather.uwyo.edu/upperair/sounding.html.

there are clouds present in this layer. Above 525 hPa, through the top of the troposphere (200 hPa, Section 14.1.2), the temperature and dew point are generally still within 3°C, indicating that clouds are likely present in that layer as well. The presence of clouds throughout the entire troposphere is not uncommon in an intense midlatitude cyclone.

In contrast, Fig. 14.2 shows a sounding from Springfield, Missouri (KSGF), after the passage of the midlatitude cyclone responsible for the 2011 central US Groundhog Day Blizzard (see Chapter 9). In the KSGF sounding, we see that the atmosphere is saturated in a small vertical layer (900–875 hPa). The dew point depression is ≤3°C from the surface to 810 hPa, indicating that clouds are likely in that layer. However, above 810 hPa, the KSGF sounding indicates considerably drier conditions, with dew point depressions >3°C. Therefore we can conclude that no clouds are present from 810 hPa to the top of the troposphere.

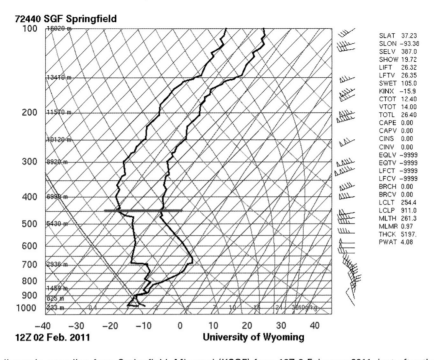

Fig. 14.2 Radiosonde sounding from Springfield, Missouri (KSGF) from 12Z 2 February 2011, just after the central US Groundhog Day Blizzard. The *thick black line* on the right (left) side represents temperature (dew point) and the *thick red line* represents the approximate level of the tropopause. Wind is plotted on the far right in kt. Reproduced with permission from the University of Wyoming Department of Atmospheric Science sounding archive, available online at: http://weather. uwyo.edu/upperair/sounding.html.

14.1.2 Tropopause Height

Another important aspect of sounding analysis is assessing the height of the tropopause. Recall that the thickness (depth) of an atmospheric layer is proportional to the mean temperature of that layer. Therefore a higher tropopause indicates a warmer (thicker) troposphere, while a lower tropopause indicates a colder (less thick) troposphere. The tropopause level can be readily identified on a sounding by remembering that temperature increases with height in the stratosphere. Therefore, the height at which a persistent and deep temperature inversion begins is the start of the stratosphere.

In Fig. 14.1, the tropopause level is located near 200 hPa, as indicated by the thick red line. This is suggestive of a warm troposphere, despite the fact the sounding in Fig. 14.1 is taken during a blizzard. Recall that the tropospheric depth is proportional to *mean* temperature, and the surface can still be below freezing when the tropopause is relatively high (Fig. 14.1). In contrast, the tropopause level in Fig. 14.2 is located near 450 hPa, indicating a much colder troposphere than in Fig. 14.1. This makes sense when one considers that the sounding in Fig. 14.2 was taken near the center of the cold upper-tropospheric trough (see Chapter 9), while the sounding in Fig. 14.1 was taken well ahead of the cold upper-tropospheric trough responsible for the Snowmageddon Blizzard. In general, the lowest tropopauses are found at the center of upper-tropospheric troughs, representative of a very cold troposphere; Section 14.2 further addresses this point.

14.1.3 Vertical Changes in Wind

In Chapter 8, we learned how to identify geostrophic temperature advection on lower-tropospheric charts. Recall that geostrophic warm-air advection (WAA) is associated with ascent, while geostrophic cold-air advection (CAA) is associated with descent. Temperature advection can be assessed from soundings by examining the changes in wind direction with increasing height.

Fig. 14.3 shows the two types of wind direction changes with height associated with temperature advection. In the first case, the wind at 850 hPa is east-southeasterly, the wind at 700 hPa is south-southeasterly, and the wind at 500 hPa is south-southwesterly. This indicates clockwise turning with increasing height, or *veering*, which is indicative of WAA. In the second case, the wind is south-southwesterly at 850 hPa, south-southeasterly at 700 hPa, and east-southeasterly at 500 hPa. This counterclockwise turning with height, or *backing*, is indicative of CAA. In the Southern Hemisphere, these associations are reversed; WAA is indicated by *backing*, while *veering* means CAA.

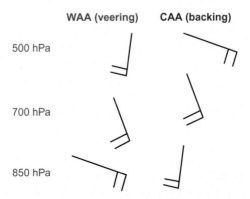

Fig. 14.3 Schematics of (left) veering and (right) backing with height wind profiles, indicative of warm-air advection (WAA) and cold-air advection (CAA), respectively.

Changes in wind speed with height can also be qualitatively suggestive of temperature advection magnitude. For example, a veering profile that also shows wind speed increasing with height suggests stronger WAA than one in which the wind speed is not changing with height (e.g., Fig. 14.3).

We can now use veering and backing to diagnose temperature advection in Figs. 14.1 and 14.2. In Fig. 14.1, the wind is easterly near the surface, southerly around 800 hPa, and west-southwesterly at 500 hPa. This indicates clockwise turning with height, i.e., *veering*, meaning that WAA is occurring in the lower troposphere. In addition, the wind speed increases by >30 kt between the surface and 500 hPa, suggesting that the WAA is strong. Since WAA is associated with ascent, it is at least partially responsible for the saturated conditions throughout the troposphere in Fig. 14.1. Although other ascent mechanisms [i.e., jet streak divergent regions, cyclonic vorticity advection (CVA)] are also important, strong WAA typically results in saturation, clouds, and precipitation, all of which were the case during the Snowmageddon event (Fig. 14.1).

In Fig. 14.2, winds are northerly around 900 hPa, turning counterclockwise with increasing height through 600 hPa, where they are westerly. This indicates CAA and sinking air in the 900–600 hPa layer. Recall that the KSGF sounding in Fig. 14.2 was taken after the passage of the surface cyclone that caused the Groundhog Day Blizzard, a scenario in which CAA is often observed (see Chapters 8 and 9). Above 900 hPa, the KSGF sounding is much less saturated than it is near the surface (Fig. 14.2), further suggestive of sinking air and adiabatic warming. In fact, the large temperature inversion in the 900–600 hPa layer is in part caused by adiabatic warming; we will explore these subsidence inversions more in the next section.

14.2 Common Types of Soundings

14.2.1 Large-Scale Patterns

Although we will cover a wide array of sounding types here, the reader should keep in mind that, like snowflakes, no two soundings are exactly alike. However, this section is intended to teach the analyst or forecaster pattern recognition skills which can be used to diagnose specific atmospheric conditions from a sounding.

Fig. 14.4 shows radiosonde soundings representative of four very different large-scale patterns. The 12Z Arctic sounding from Baker Lake, Northwest Territories, Canada is characterized by

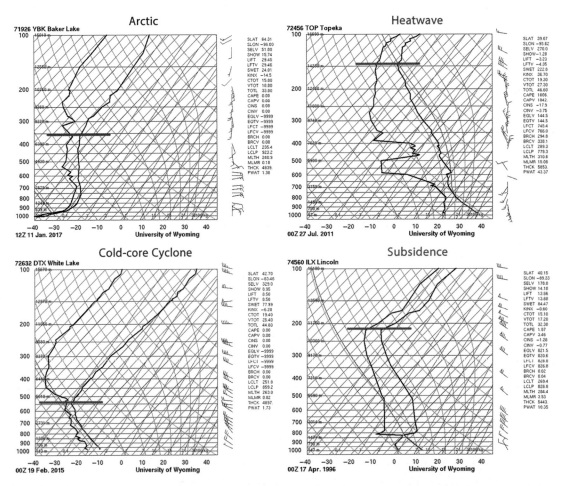

Fig. 14.4 Example soundings for (top left) Arctic [Baker Lake, Northwest Territories (CYBK), 12Z 11 January 2017], (top right) Heat wave [Topeka, Kansas (KTOP), 00Z 27 July 2011], (bottom left) Cold-core Cyclone [White Lake, Michigan (KDTX), 00Z 19 February 2015], and (bottom right) Subsidence [Lincoln, Illinois (KILX), 00Z 17 April 1996]. In each panel, the *thick black line* on the right (left) side represents temperature (dew point) and the *thick red line* represents the approximate level of the tropopause. Wind is plotted on the far right in kt. Reproduced with permission from the University of Wyoming Department of Atmospheric Science sounding archive, available online at: http://weather.uwyo.edu/upperair/sounding.html.

extremely cold surface temperatures near −40°C (−40°F), a dramatic temperature inversion just above the surface and a low tropopause height of approximately 350 hPa. Recall that the coldest surface air is usually found near the center of Arctic surface anticyclones, where winds are light and skies are generally clear, promoting strong radiational cooling. That is certainly the case at Baker Lake, where winds are quite weak below the radiation inversion.

The 00Z 27 July 2011 Topeka, Kansas sounding in Fig. 14.4 is a prime example of a heat wave, taken during the long-duration central US heat wave that year. The atmosphere is characterized by a very high tropopause (150 hPa), extremely hot surface temperatures near 35°C (95°F), and relatively weak winds throughout the entire troposphere. Both the high tropopause height and hot surface temperatures exemplify tropospheric warmth, while the weak winds correspond to the fact that most heat waves occur in the presence of upper-tropospheric ridges. In addition, the entire troposphere is relatively dry, evidence of the large-scale subsidence (sinking air) typically seen during a heat wave.

The Cold-core Cyclone sounding at White Lake (Detroit), Michigan was taken from the center of an upper-tropospheric trough during a cold-air outbreak in February 2015. Cold-core Cyclone soundings are characterized by very low tropopause heights (550 hPa in Fig. 14.4), representative of an extremely dense and cold troposphere. The Cold-core Cyclone sounding is also relatively saturated, which is typical near upper-tropospheric troughs. Finally, the tropopause height is actually lower in the Cold-core Cyclone sounding than in the Arctic sounding, indicative of how dominant surface radiative cooling is in the Arctic; this is a good example of how surface temperatures are not completely representative of the mean tropospheric temperature.

The last sounding in Fig. 14.4 is labeled Subsidence, and exemplifies a situation where there is a lot of sinking air in the anticyclonic vorticity advection (AVA) region ahead of an upper-tropospheric ridge. The key attribute of a Subsidence sounding is the large inversion in the mid-troposphere (near 800 hPa in Fig. 14.4); recall that as air descends, it warms adiabatically and dries, creating a larger dew point depression at the bottom of the subsidence layer than at the top. Although CAA is a large-scale mechanism associated with descent, no backing with height is observed in Fig. 14.4. However, in Fig. 14.2, where strong CAA is present, there is a pronounced subsidence inversion. Because AVA and CAA are both associated with large-scale descent, the forecaster can infer that Subsidence soundings that do not feature backing wind profiles are predominantly associated with AVA.

In Fig. 14.5, soundings are shown from two different Atlantic tropical cyclones (TCs), Katrina (2005) and Matthew (2016). Typical characteristics of TC soundings include environmental lapse rates that are moist adiabatic for the entire depth of the troposphere and exceptionally high tropopause heights. In both soundings, the whole troposphere is saturated and the tropopause levels are near 100 hPa, depicting very moist and warm air columns, respectively. Another aspect of TC soundings is that the fastest

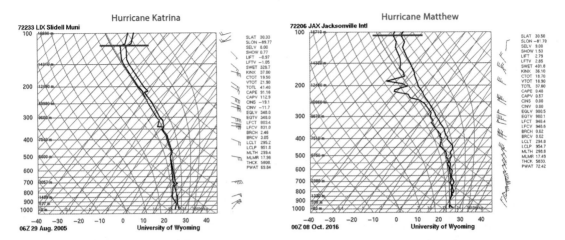

Fig. 14.5 Example tropical cyclone soundings for (left) Hurricane Katrina [Slidell, Louisiana (KLIX), 06Z 29 August 2005], and (right) Hurricane Matthew [Jacksonville, Florida (KJAX), 00Z 8 October 2016]. In each panel, the *thick black line* on the right (left) side represents temperature (dew point) and the *thick red line* represents the approximate level of the tropopause. Wind is plotted on the far right in kt. Reproduced with permission from the University of Wyoming Department of Atmospheric Science sounding archive, available online at: http://weather.uwyo.edu/upperair/sounding.html.

wind speeds are always found in the lower troposphere, because TCs are warm-core cyclones that weaken with increasing height. In the Katrina and Matthew soundings, the fastest wind speeds are below 800 hPa, indicative of a solid warm-core cyclone. Finally, note that the surface wind speeds in the two TC soundings do not match the peak TC intensity in either case, specifically Category 3 (at landfall) Katrina. This exemplifies how limited temporal and spatial resolution of radiosonde soundings can cause issues, as it is unlikely that a radiosonde will be launched in the TC eyewall (region of peak winds) at the exact time of landfall or closest passage.

14.2.2 Frontal

Fig. 14.6 illustrates examples of Warm and Cold Front soundings, from Maniwaki, Quebec and Lake Charles, Louisiana, respectively. Because fronts can occur in any season, temperatures in frontal soundings can vary widely based on location and season. However, there are still common signatures that the analyst or forecaster should be able to recognize. The Maniwaki Warm Front sounding in Fig. 14.6 is characterized by cold surface temperatures, a large temperature inversion just above the surface, and a saturated troposphere. One of the strongest indicators of a warm front aside from the "nose" of warm air (i.e., the

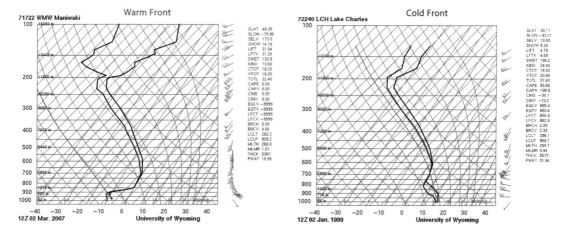

Fig. 14.6 Example soundings for (left) Warm Front [Maniwaki, Quebec (CWMW), 12Z 2 March 2007], and (right) Cold Front [Lake Charles, Louisiana (KLCH), 12Z 2 October 1999]. In each panel, the *thick black line* on the right (left) side represents temperature (dew point). Wind is plotted on the far right in kt. Reproduced with permission from the University of Wyoming Department of Atmospheric Science sounding archive, available online at: http://weather.uwyo.edu/upperair/sounding.html.

temperature inversion) is a wind profile veering with height, indicative of WAA. WAA is associated with ascent and thus clouds and precipitation, exemplified by the saturated troposphere at Maniwaki. The WAA and "warm nose" layers almost always coincide, and as we will see in the next section, can frequently result in ice pellets and freezing rain during the cold season.

The Cold Front sounding from Lake Charles (Fig. 14.6) similarly shows a mostly saturated troposphere. However, the winds are strongly backing in the 750–600-hPa layer, indicative of CAA. Because most cold fronts are narrow, the amount and depth of saturation in cold front soundings can vary widely; sometimes a subsidence inversion will be more prominent than in the example in Fig. 14.6, particularly just below the CAA layer. As with all sounding types, it is important for the forecaster to supplement their frontal analysis by examining surface observations taken at the same time.

14.2.3 Frozen Precipitation

Frozen and mixed precipitation are some of the most difficult things to forecast for in meteorology. In such instances, there are often many competing large- and small-scale processes such as WAA, evaporation, melting, and adiabatic cooling. Recall that the overwhelming majority of midlatitude precipitation starts as snow or ice (hail) at high altitudes. In the winter, what happens to the snow as it falls is entirely dependent on the tropospheric temperature profile.

Fig. 14.7 shows example soundings for snow, ice pellet (sleet), and freezing rain events, respectively. In all three soundings, the lower troposphere is saturated and the surface temperature is below freezing. The Snow sounding from Bismarck, North Dakota features the entire troposphere below freezing, with the bold pink line highlighting the 0°C isotherm. A warm—but still subfreezing—nose is seen at approximately 700 hPa, where veering (WAA) is the most pronounced. The Ice Pellet (sleet) sounding from central Illinois is similar to the Bismarck sounding with respect to surface temperatures, lower-tropospheric saturation, and veering winds. The key difference is found in the 850–800-hPa layer, where the temperature is slightly above freezing. Recall that ice pellets are melted snowflakes that partially refreeze before hitting the ground. Ice pellet soundings are characterized by shallow above-freezing layers with temperatures typically reaching only 1–2°C.

The final sounding in Fig. 14.7 is a Freezing Rain sounding taken during an ice storm in Kansas and Oklahoma in January 2017. While the surface temperature is barely below freezing, there is a large temperature inversion and extremely deep and pronounced above-freezing warm nose in the 950–700-hPa layer. The depth and magnitude of the above-freezing layer far exceed those in the Ice Pellet sounding. Recall that freezing rain is snow that melts completely and only refreezes upon contact with the surface. Another typical element of freezing rain soundings is that the surface winds are completely "separated" from the winds slightly aloft. The Norman sounding in Fig. 14.7 demonstrates this signature: surface winds are coming from the northeast, helping to replenish the supply of surface cold air, while winds starting just above the surface are from the southwest, advecting in warmer air aloft.

It is important that the forecaster realize that frozen and mixed precipitation situations can evolve quickly. Temperature advection, adiabatic warming/cooling (i.e., ascent/descent), and diabatic warming/cooling (i.e., melting, evaporation, etc.) are often competing processes, which make it difficult for the atmosphere to maintain a specific temperature profile for a long period of time. Numerical weather prediction models frequently struggle with frozen precipitation forecasts, so it is vital that the forecaster utilize observations and physical reasoning for the duration of the event.

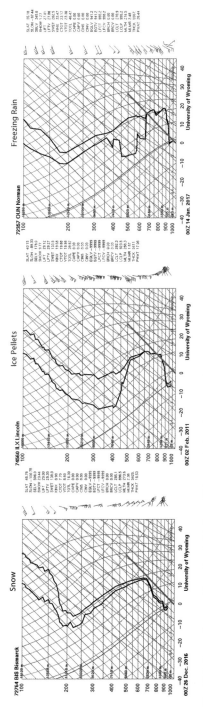

Fig. 14.7 Example soundings for (left) Snow [Bismarck, North Dakota (KBIS), 00Z 26 December 2016], (middle) Ice Pellets [Lincoln, Illinois (KILX), 00Z 2 February 2011], and (right) Freezing Rain [Norman, Oklahoma (KOUN), 00Z 14 January 2017]. In each panel, the *thick black line* on the right (left) side represents temperature (dew point), and the *thick pink line* highlights the 0°C isotherm. Wind is plotted on the far right in kt. Reproduced with permission from the University of Wyoming Department of Atmospheric Science sounding archive, available online at: http://weather.uwyo.edu/upperair/sounding.html.

14.2.4   Influences of Terrain

In regions of elevated or complex terrain, forecasting can be exponentially more challenging than in flat areas. It is vital to possess knowledge of terrain location, peak height, and localized effects. Soundings can be an extremely useful analysis and forecasting tool in regions where terrain is a primary factor in producing ascent or descent. In this section, we will focus on Denver, Colorado, located in the lee of the Rocky Mountains. The axis of the Rocky Mountains is generally along a southeast-northwest line, with Denver located just to the east of the highest peaks. Typical lower-tropospheric upslope flow in Denver ranges from northeasterly to southeasterly, while typical downslope flow ranges from northwesterly to southwesterly. In Fig. 14.8, the Upslope sounding shows lower-tropospheric northeasterly flow, which is ascending the terrain perpendicular to the Rockies. As a result of upslope ascent, the troposphere is saturated, with clouds and precipitation likely, as evidenced by moist-adiabatic environmental lapse rates throughout most of the troposphere. In addition, the winds in Fig. 14.8 are veering with height, indicating WAA, another ascent mechanism. Because temperatures are below freezing in the entire troposphere, we can conclude that snow is likely falling, and indeed this sounding was taken during a high-impact Denver area snowstorm in December 2006.

The Downslope sounding in Fig. 14.8 depicts a completely different weather situation from the Upslope sounding. Winds are

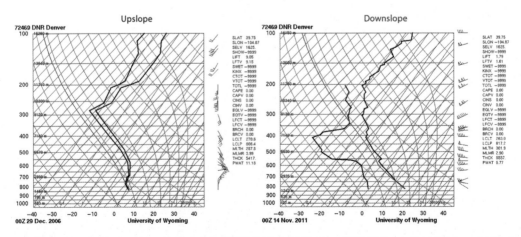

Fig. 14.8 Example soundings at Denver, Colorado (KDNR) for (left) Upslope (00Z 29 December 2006), and (right) Downslope (00Z 14 November 2011). In each panel, the *thick black line* on the right (left) side represents temperature (dew point). Wind is plotted on the far right in kt. Reproduced with permission from the University of Wyoming Department of Atmospheric Science sounding archive, available online at: http://weather.uwyo.edu/upperair/sounding.html.

generally westerly throughout the entire troposphere, indicating that air is descending the Rockies as it blows toward Denver. As a result, the troposphere is very dry and environmental lapse rates from the surface to 600 hPa are essentially dry adiabatic, evidence of adiabatic warming. Fig. 14.8 shows that at a location like Denver, it is very difficult to get clouds and precipitation on a downslope wind, and similarly challenging to get clear skies and warm weather on an upslope wind. Therefore, knowing the location and characteristics of the topography in your forecast region is crucial to being able to accurately forecast the weather. Furthermore, geographical knowledge is vital; for example, while the Rockies are located to the west of Denver and the Appalachians to the west of Washington, District of Columbia, the highest terrain in California is located to the east or northeast of the big cities. For example, in Los Angeles, the strongest downsloping occurs on a northeasterly wind.

14.2.5 Convective

Thunderstorms are one of the biggest meteorological hazards to life and property, including the transportation industry. Because they are a mesoscale phenomenon, forecasters can use an ingredient-based methodology to predict environments favorable for thunderstorms. To that end, soundings are one of the most powerful analysis and forecasting tools for convection. Recall that the three necessary ingredients for thunderstorms are ascent (lift/trigger), moisture, and instability. A fourth ingredient, vertical wind shear, is necessary to make storms last longer and be more intense. Using soundings, we can essentially diagnose whether the values of each ingredient are sufficient to produce a thunderstorm, and if so, how severe the convection may potentially be. In this section, we will examine two sets of convective soundings. The first contains two soundings, one before and one during an EF-5 tornado in Oklahoma, while the second compares and contrasts soundings for three types of convective hazards: tornado, microburst, and bow echo/derecho.

Fig. 14.9 focuses on the atmospheric environment before and during the 20 May 2013 Moore, Oklahoma EF-5 tornado,[2] which caused immense damage to life and property just south of Oklahoma City. Both soundings are from nearby Norman, Oklahoma (KOUN), one from 12Z, taken approximately 8–9 h before the

[2]A summary of the 2013 Moore, Oklahoma EF-5 tornado is available from the National Weather Service (NWS) Norman, Oklahoma, online at: https://www.weather.gov/oun/events-20130520.

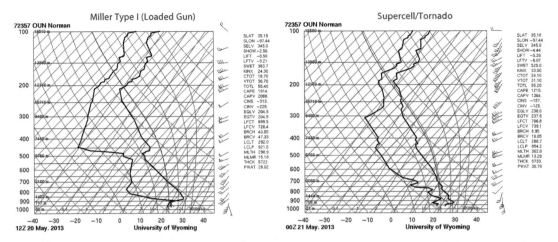

Fig. 14.9 Example soundings at Norman, Oklahoma (KOUN) for (left) Miller Type I (Loaded Gun, 12Z 20 May 2013), and (right) Supercell/Tornado (00Z 21 May 2013). In each panel, the *thick black line* on the right (left) side represents temperature (dew point), wind (kt) is plotted on the right side, and convective parameters are listed on the far right. Reproduced with permission from the University of Wyoming Department of Atmospheric Science sounding archive, available online at: http://weather.uwyo.edu/upperair/sounding.html.

tornado, and one from 00Z, 2–3 h after the tornado occurred. The 12Z sounding is what is typically known as a Miller Type I or Loaded Gun sounding, and is commonly seen in 12Z morning soundings before a tornadic event. At the surface, the temperature and dew point are nearly equal, suggesting good low-level moisture. Meanwhile, a large capping inversion ("cap") is present in the 900–850-hPa layer, which must be overcome later in the day if storms are going to form. Above the capping inversion, environmental lapse rates are very steep (near dry adiabatic); this elevated mixed layer (EML) suggests that if the cap breaks later in the day, CAPE will be large and updrafts will be strong. Finally, the surface–500-hPa (~0–6-km) vertical shear is strong, as the winds veer from 5 kt southeasterly at the surface to 50 kt west-southwesterly at 500 hPa.

The Loaded Gun name comes from the goalpost ("gun") appearance of the lower-troposphere in the 12Z sounding, and also from the fact that the atmosphere is primed for intense convection if certain processes happen during the day. For the Loaded Gun to be fired, the cap needs to break. This is typically accomplished by some combination of surface heating and moistening, and mid-tropospheric cooling. In addition, there needs to be an ascent (trigger) mechanism such as CVA associated with an upper-tropospheric trough. In the case of the Moore EF-5, by mid-afternoon, the surface had heated to near 30°C (86°F), and the mid-troposphere cooled substantially. This enabled large values of CAPE to develop, and thunderstorms to fire along a dry-line once the CVA from the associated upper-tropospheric trough moved into Oklahoma. The Supercell/Tornado sounding in Fig. 14.8 also shows a strong veering profile, suggesting WAA and very strong vertical shear across central Oklahoma. WAA provides an additional ascent mechanism, while veering vertical shear allows storms to rotate and helps them become potentially tornadic.

Soundings are also a powerful tool to differentiate between types of convective situations. In Fig. 14.10, three such situations are presented. The Tornado sounding was taken from Lamont, Oklahoma at 00Z 5 May 2007, when an EF-5 tornado was occurring just to the northwest in Greensburg, Kansas.[3] The 00Z Lamont sounding shows steep mid-tropospheric lapse rates associated with extreme instability (CAPE >3500 J kg^{-1}), strong veering wind shear, and high surface dew points (moisture). Both WAA and CVA ahead of an upper-tropospheric trough (not shown) provided the ascent.

The Microburst sounding in Fig. 14.10 has similarly large (>2500 J kg^{-1}) CAPE values and steep mid-tropospheric lapse rates but much weaker vertical shear. In addition, while surface dew points are relatively high, the surface temperature is extremely warm, forming below 850 hPa what forecasters refer to as an "inverted V" signature on a sounding. This signature is most typically associated with downburst potential, and in fact, the Microburst sounding was taken at Stephenville, Texas (KSEP), just south of Dallas-Fort Worth, near the time of the crash of Delta Flight 191[4] at Dallas-Fort Worth International Airport. The crash was later determined to be due to a microburst in a non-severe thunderstorm, and later spurred the development of Doppler Radar technology that has since prevented major airline accidents due to microbursts. Note that microbursts can occur in any thunderstorm, as evidenced by the weak vertical wind shear in Fig. 14.10, provided that ascent, moisture, and instability are sufficient. The relatively dry layer ("inverted V") near the surface is a crucial aspect of pattern recognition for such cases.

---

[3] A summary of the 2007 Greensburg, Kansas EF-5 tornado is available from the NWS Dodge City, Kansas, online at: https://www.weather.gov/media/ddc/Greensburg_1year_later.pdf.
[4] A summary of the Flight 191 crash and the microburst that caused it is available from the Federal Aviation Administration (FAA), online at: http://lessonslearned.faa.gov/ll_main.cfm?TabID=1&LLID=32&LLTypeID=2.

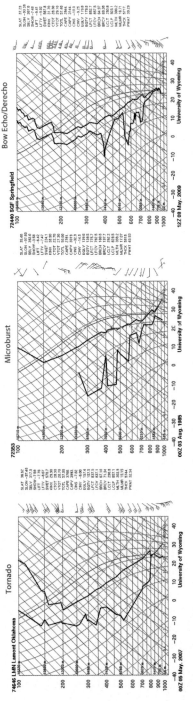

Fig. 14.10 Example soundings for (left) Tornado [Lamont, Oklahoma (KLMN), 00Z 5 May 2007], (middle) Microburst [Stephenville, Texas (KSEP), 00Z 3 August 1985], and (right) Bow Echo/Derecho [Springfield, Missouri (KSGF), 12Z 8 May 2009]. In each panel, the *thick black line* on the right side represents temperature (dew point), wind (kt) is plotted on the right side, and convective parameters are listed on the far right. Reproduced with permission from the University of Wyoming Department of Atmospheric Science sounding archive, available online at: http://weather.uwyo.edu/upperair/sounding.html.

The final sounding in Fig. 14.10 is from an infamous derecho that impacted the Great Lakes and Northeastern United States in 1995.[5] Like the Tornado and Microburst soundings, the Bow Echo/Derecho sounding features a large amount of CAPE and a pronounced EML. The key differentiating sounding features for bow echoes, however, are the extremely strong winds in the mid-troposphere and the fact that the vertical shear is predominantly speed shear (not directional). Unlike supercells and tornadoes, squall lines and derechos occur preferentially when strong vertical wind shear is unidirectional instead of rotating (i.e., veering) with height.

14.3 Questions and Practice Exercises

1. Explain why the tropopause height is higher in a TC sounding than in a cold-core low sounding.
2. The following upper-air winds are reported from the radiosonde sounding at Stony Plain, Alberta (near Edmonton):
   * 700 hPa: southerly at 12 kt.
   * 600 hPa: west-southwesterly at 21 kt.
   * 500 hPa: northwesterly at 34 kt .

   Identify whether the wind is veering or backing with height and explain which type of temperature advection must be occurring.
3. Explain the differences between a typical ice pellet sounding and a typical freezing rain sounding.
4. Explain the important differences between a typical microburst sounding and a typical tornado sounding.
5. Below (Fig. 14.11) are three radiosonde soundings from unidentified locations at unidentified times. Match each sounding with *one* of the following sounding types:
   * Cold front.
   * Heat wave.
   * Arctic.

   Justify your answers with physical reasoning.
6. Fig. 14.12 shows a MSLP (hPa, solid black contours) and 1000–500-hPa thickness (dam, dashed red contours) chart from some day at some time, with Points A and B marked in bold. Match each point with one of the three soundings in Question 5 (Fig. 14.11). Only two of the three soundings should be chosen. Justify your answers using physical reasoning.

---

[5]A summary of the July 1995 derechoes is available from the NWS Storm Prediction Center (SPC), online at: http://www.spc.noaa.gov/misc/AbtDerechos/casepages/jul1995derechopage.htm.

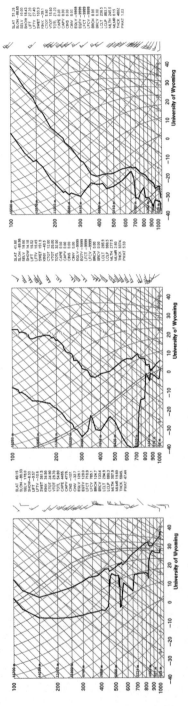

Fig. 14.11 The three soundings required for Question 5. In each panel, the thick black line on the right (left) side represents temperature (dew point), wind (kt) is plotted on the right side, and convective parameters are listed on the far right. Reproduced with permission from the University of Wyoming Department of Atmospheric Science sounding archive, available online at: http://weather.uwyo.edu/upperair/sounding.html.

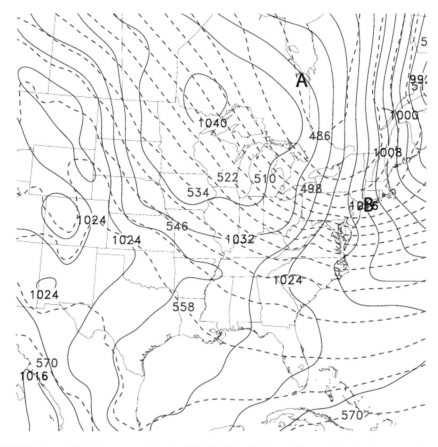

Fig. 14.12 MSLP (hPa, *solid black contours*) and 1000–500-hPa thickness (dam, *dashed red contours*) to be used for Question 6. Points A and B are *bolded in blue*.

15

WEATHER FORECASTING BASICS

Predicting the future is not an easy task. Fortunately, with the help of modern computing power, numerical weather prediction (NWP) model skill has increased and continues to increase exponentially over time. However, the role of the human weather forecaster should not be limited to reading raw computer model output. The forecaster, through knowledge of underlying atmospheric processes and pattern recognition, should be able to make surface temperature and precipitation forecasts that are as skillful as, or more accurate than, than raw NWP output, especially during anomalous and extreme weather events. In this chapter, we discuss simplistic, yet effective, introductory temperature and precipitation forecast techniques, with the understanding that increased human forecast skill comes only through practice and repetition.

15.1 Temperature Forecasting

15.1.1 Maximum Temperature

In this section, we are going to use NWP model forecast upper-air data and our knowledge of the appropriate physical mechanisms to make a surface maximum temperature forecast. This method is primarily based on the concept of lapse rates and surface processes that act to warm or cool the surface. Because NWP models exhibit greater skill (accuracy) for upper-air variables than for surface temperatures, mainly due to necessary parameterizations of surface fluxes, the human forecaster can add value to NWP model output using the approach presented here.

The forecast example used throughout this section will be for Davenport, Iowa (KDVN) on a June afternoon of some year. KDVN is located at an elevation of 230 m (0.230 km) above sea level. To start, we need to know the model forecast 850-hPa temperature

Synoptic Analysis and Forecasting. https://doi.org/10.1016/B978-0-12-809247-7.00015-6

and 850-hPa geopotential height for mid-afternoon, the warmest time of day. In this case, our favorite NWP model is predicting an 850-hPa temperature of 18°C and an 850-hPa geopotential height of 1470 m (1.47 km). To then calculate the forecast maximum surface temperature at KDVN, use the following steps:

1. Calculate the forecast sea-level temperature, assuming that the environmental lapse rate (ELR) is equal to the dry adiabatic lapse rate (DALR, 9.8°C km^{-1}). The ELR typically cannot exceed the DALR, except in very shallow layers.

    Sea-level temperature = 850-hPa temperature + DALR * 850-hPa Geopotential Height.

    At KDVN: Sea-level temperature = 18°C + 9.8°C km^{-1} * 1.47 km = 32.4°C.

    Note: The 850-hPa geopotential height is written in km to match the units of the DALR.

2. Adjust the calculated sea-level temperature for the station elevation, which in the case of KDVN is 230 m. The resulting temperature is called the "max max," or the surface maximum temperature that will occur if the lower-tropospheric ELR is dry adiabatic between 850 hPa and the surface.

    "Max max" temperature = Sea-level temperature − DALR * Station Elevation.

    At KDVN: "Max max" temperature = 32.4°C − 9.8°C km^{-1} * 0.230 km = 30.15°C.

3. Now that you have calculated the forecast "max max" temperature, you will need to adjust it based on your knowledge of various weather factors such as cloud cover, wind speed and direction, terrain, etc. This is done using the checklist in Table 15.1. For the KDVN example (Table 15.1), we conclude after following the checklist that 1.5°C should be added to 30.15°C, which gives us a forecast maximum surface temperature of 31.65°C, or ~90°F.

This methodology for forecasting maximum temperatures may sometimes seem like magic, but it can often give the human forecaster an advantage over NWP model guidance. In the KDVN example, two different NWP model guidance forecasts from the previous evening's 00Z runs had 87°F as the forecast maximum temperature, while the National Weather Service forecast 92°F. The actual maximum temperature at KDVN verified at 90°F, justifying our methodology and slight deviation from NWP guidance. In short, we have just provided added value to NWP forecasts!

There are a few other issues to consider when using the checklist to forecast maximum temperatures:

- Specific upslope and downslope wind directions vary by location. For example, a downslope wind in Denver is westerly, but

Table 15.1 Step-By-Step Checklist Used to Adjust the "Max Max" Temperature to Calculate Forecast Surface Maximum Temperature, Integrating Various Important Weather Factors

Step	Atmospheric Condition and Impact on Temperature	Result for KDVN Example
(i)	**Cloudy skies?** *If no, skip to (ii).* If you expect completely cloudy conditions all day, adjust the "max max" temperature by −5°C. Adjust by −2 to −3°C for broken cloud cover *Go to (ii)*	No clouds are forecast. *Skip to (ii)*
(ii)	**Sunshine through cirrus?** *If no, skip to (iii).* Depending on the duration and sky coverage of the cirrus clouds, adjust by −1 to −3°C. *Go to (iii)*	No cirrus clouds are forecast. *Skip to (iii)*
(iii)	**Snow cover?** *If no, skip to (iv)* Snow, especially fresh snow, has a high albedo and will prevent surface absorption of incoming solar radiation. Adjust by −2 to −3°C. *Go to (iv)*	There is no snow cover present, as it is June. *Skip to (iv)*
(iv)	**Steady precipitation all day?** *If no, skip to (v)* Evaporational cooling from falling rain or snow will reduce the temperature. Adjust by −3°C. *Go to (v)* Note: Another way to calculate the surface temperature if it is raining or snowing heavily for a long period of time is to simply calculate the surface **wet bulb temperature** T_w, which is the minimum temperature you can have if all the water is evaporated into the atmosphere. See Chapter 13	No precipitation is forecast. *Skip to (v)*
(v)	**Light winds (≤5 kt)?** *If no, skip to (vi).* Both sunshine and wind are required to break morning inversions. Particularly in the cool season, this can leave the maximum temperature well short of the "max max," despite sunshine. Depending on the strength of the morning inversion, adjust by −2 to −4°C. *Go to (vi)*	Surface wind speeds are forecast to be 20–25 kt. *Skip to (vi)*
(vi)	**Upslope winds?** *If no, skip to (vii).* The ELR tends to be more stable in the presence of upslope flow. Adjust by −3°C. *Go to (vii)*	The surface wind direction is forecast to be from the south-southwest, which is not upslope. *Skip to (vii)*

Continued

Table 15.1 Step-By-Step Checklist Used to Adjust the "Max Max" Temperature to Calculate Forecast Surface Maximum Temperature, Integrating Various Important Weather Factors—cont'd

Step	Atmospheric Condition and Impact on Temperature	Result for KDVN Example
(vii)	**Strong winds from a large warm body of water?** *If no, skip to (viii).* These winds can help bring in warm moist air (e.g., south winds from the Gulf of Mexico), which can result in larger dew points and a modification of the air mass. Adjust by +1 to +2°C. *Go to (vii)*	Strong winds from the south are forecast. *Adjust by + 1.5°C*
(viii)	**Downslope winds?** These winds can be quite strong, mixing a deep layer of air from terrain peak height, creating a situation where the ELR equals the DALR well *above* 850 hPa. This results in a warmer "max max" than would otherwise occur. Adjust by +1 or +2°C in the cool season and +3°C in the warm season	Winds are not forecast to be westerly and thus are not downslope. *Do not adjust*

A general description is included in the middle column, while specific results for the KDVN example presented in this chapter are shown in the right-hand column.

a downslope wind in San Diego is from the northeast. The forecaster should always be aware of where the higher terrain is located relative to the forecast region.

- At high-elevation locations (e.g., Denver), the 850-hPa pressure surface is often located below the ground. Therefore, to calculate the "max max" temperature, the best course of action is to instead use the 700-hPa temperature and geopotential height.

Finally, the mathematical calculation of the "max max" temperature can be alternatively completed using a Skew-T diagram, assuming that the forecaster knows the correct upper-air temperature and station elevation. Just start at the appropriate upper-air (e.g., 850-hPa for KDVN) temperature and move to the station surface elevation along the DALR, then read the temperature at that height, which is your "max max." Subsequently, the forecaster can use the checklist in Table 15.1 to adjust the "max max" as needed.

15.1.2 Minimum Temperature

In this section, we are going to use the afternoon maximum temperature (observed or forecast) and our knowledge of the appropriate physical mechanisms to make a surface minimum temperature forecast for the subsequent night. As with the maximum temperature forecast procedure, this method is primarily based on the concept of lapse rates and surface processes that act to warm or cool the surface temperature.

The forecast example used throughout this section is for Oklahoma City, Oklahoma (KOKC) on a June evening of some year. KOKC is located at an elevation of 390 m (0.390 km) above sea level. To start, we need to know the observed or forecast maximum temperature for that day. In this case, Oklahoma City reported a daily maximum surface temperature of 32.2°C (90°F). To then calculate the forecast minimum surface temperature at KOKC, use the nearby Norman, Oklahoma (KOUN) 0000Z sounding in Fig. 15.1 and proceed with the following steps:

1. Move dry adiabatically from the KOKC maximum surface temperature (32.2°C) up to 850 hPa (bold red line in Fig. 15.1).
2. Return to the surface elevation moist adiabatically (bold green line in Fig. 15.1) from the 850-hPa temperature that you calculated in (1). This is your "max min" temperature, or the warmest that we can reasonably expect the surface minimum temperature to be that night.

   At KOKC: "Max min" temperature = 21.5°C (Fig. 15.1).
3. Now that you have calculated the forecast "max min" temperature, you will need to adjust it based on your knowledge of various weather factors such as cloud cover, wind speed and direction, terrain, etc. This can be done using the checklist in Table 15.2. For the KOKC example (Table 15.2), we conclude that 1°C should be added to 21.5°C (Table 15.2), which gives us a forecast minimum surface temperature of 22.5°C, or ~72.5°F.

In the KOKC example, two different NWP model guidance forecasts from the previous morning's 1200Z runs predicted 70°F as the forecast minimum temperature, while the National Weather Service forecast 71°F. The actual maximum temperature at KOKC ended up verifying at 73°F, justifying our methodology and slight deviation from NWP guidance. Once again, we have added value to NWP forecasts.

Finally, it is important to note that unlike the "max-max," the "max-min" should generally be calculated using a Skew-T diagram, because the moist-adiabatic lapse rate (MALR) is not constant, and therefore difficult to calculate mathematically. One could hypothetically use an average value of the MALR for a given temperature range, but doing so could result in calculation error.

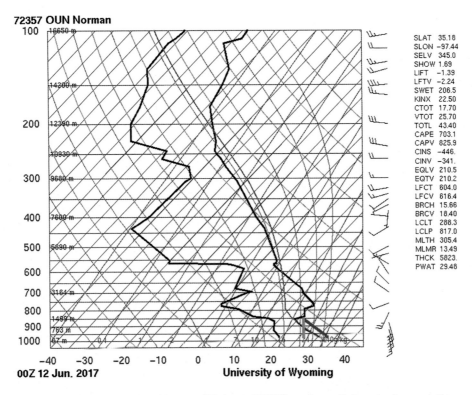

Fig. 15.1 00Z radiosonde sounding from Norman, Oklahoma (KOUN) used to calculate the "max min" temperature at Oklahoma City (KOKC). The *bold red line* represents moving dry adiabatically from the daily surface maximum temperature to 850 hPa, and the *bold green line* represents moving along the moist adiabat from 850 hPa to the surface elevation, in order to retrieve the "max min" temperature. Reproduced with permission from the University of Wyoming Department of Atmospheric Science sounding archive, available online at: http://weather.uwyo.edu/upperair/sounding.html.

15.2 Precipitation Forecasting

15.2.1 Probability of Precipitation

In Chapter 9, we outlined physical mechanisms associated with vertical motion. Recall that ascent is associated with clouds and precipitation, while descent results in clear skies and pleasant weather. As in Table 9.1, Table 15.3 outlines the sign of each mechanism associated with ascent and descent, and the last row incorporates the role of flow over terrain, as applicable. In this section, we will learn how to create basic probability of precipitation (PoP) forecasts based on climatology and an assessment of physical mechanisms associated with vertical motion.

We will use the assumption that the climatological PoP in most locations in North America east of the Rockies is ~30%. Table 15.4 contains recommended adjustments to the climatological PoP

Table 15.2 Step-By-Step Checklist Used to Adjust the "Max Min" Temperature to Calculate Forecast Surface Minimum Temperature, Integrating Various Important Weather Factors

Step	Atmospheric Condition and Impact on Temperature	Result for KOKC Example
(i)	Clear skies? *If no, skip to (ii).* Clear skies allow for radiational cooling, even on a breezy night. Adjust the "max min" temperature by −3°C. *Go to (ii)*	No clouds are forecast. *Adjust by −3°C*
(ii)	Windy (≥25 kt) at 850 or 700 hPa? *If no, skip to (iii).* Strong winds aloft may mix down to the surface, keeping the surface temperature warmer than it otherwise would be. Adjust by +2°C. *Go to (iii)*	Winds at 850 hPa are forecast to be 33 knots. *Adjust by +2°C*
(iii)	Snow cover? *If no, skip to (iv).* Snow, especially fresh snow, radiates infrared radiation away from the surface very efficiently. Adjust by −2°C to −3°C for fresh and deep snow cover. *Go to (iv)*	There is no snow cover present, as it is June. *Skip to (iv)*
(iv)	Steady precipitation for at least part of the night? *If no, skip to (v).* Evaporational cooling from falling rain or snow will reduce the air temperature. Adjust by −3°C. *Go to (v)* Note: The surface wet-bulb temperature method discussed in Table 15.1 still works at night, assuming steady precipitation is falling.	No precipitation is forecast. *Skip to (v)*
(v)	Light surface winds (≤5 knots)? *If no, skip to (vi).* Even on a cloudy night, there will be some heat loss if the winds are calm or very light. Adjust by −1 to −2°C. *Go to (vi)*	Surface wind speeds are forecast to be 10–15 kt. *Skip to (vi)*
(vi)	Upslope winds or winds from a cold air source region? *If no, skip to (vii).* Wind from a cold air source region tends to advect surface cold air. Upslope flow adiabatically cools the temperature. Adjust by −3 to −5°C, depending on the strength of the wind. *Go to (vii)*	Surface wind direction is forecast to be from the south, which is not upslope or from a cold air source region. *Skip to (vii)*

Continued

Table 15.2 Step-By-Step Checklist Used to Adjust the "Max Min" Temperature to Calculate Forecast Surface Minimum Temperature, Integrating Various Important Weather Factors—cont'd

Step	Atmospheric Condition and Impact on Temperature	Result for KOKC Example
(vii)	**Winds from a large body of water?** *If no, skip to (viii).* Large bodies of water can moderate nighttime temperature. In addition, wind from a large, warm body of water can help bring in warm moist air (e.g., south winds from the Gulf of Mexico), which can result in larger dew points and a modification of the air mass. Adjust by +1 to +4°C, depending on proximity to the body of water and strength of the wind. *Go to (viii)*	Strong winds from the south are forecast, but the Gulf of Mexico is relatively far away from KOKC. *Adjust by + 1°C*
(viii)	**Downslope winds or winds from a warm air source?** Downslope winds can mix a deep layer of air down from terrain peaks, while winds from a warm direction can advect in warmer surface air. Adjust by +1 or +2°C in the cool season and +2 to +4°C in the warm season, depending on the strength of the wind	Winds are not forecast to be downslope, but are from a warm source region. *Adjust by + 1°C*

A general description is included in the middle column, while specific results for the KOKC example presented in this chapter are shown in the right-hand column.

Table 15.3 Overview of the Mechanisms Associated With Vertical Motion, Adapted for This Chapter From Table 9.1

	Ascent	Descent
Jet streak	Divergent region	Convergent region
Upper-tropospheric vorticity advection	Cyclonic vorticity advection (CVA)	Anticyclonic vorticity advection (AVA)
Lower-tropospheric temperature advection	Warm-air advection (WAA)	Cold-air advection (CAA)
Flow over terrain	Upslope	Downslope

Table 15.4 Adjustments to PoP for Weak (Middle Column) and Strong (Right-Hand Column) Magnitudes of Each Physical Mechanism for Vertical Motion (Left-Hand Column)

	PoP Adjustment for Weak (%)	PoP Adjustment for Strong (%)
Divergent jet region	+10	+20
Convergent jet region	−10	−20
Upper-tropospheric CVA	+10	+20
Upper-tropospheric AVA	−10	−20
Lower-tropospheric WAA	+10	+20
Lower-tropospheric CAA	−10	−20
Upslope flow	+15	+30
Downslope flow	−15	−30

based on an assessment of physical mechanisms associated with vertical motion and the relative intensity of each. The forecast examples presented in this section are from the 2011 Groundhog Day Blizzard, discussed in Chapter 9, for four different cities: Chicago, Illinois; Fort Smith, Arkansas; Pittsburgh, Pennsylvania; and Greensboro, North Carolina (Fig. 15.2).

As discussed in Chapter 9, Chicago is located in a region of dual jet divergence, strong upper-tropospheric CVA, and strong lower-tropospheric WAA (Fig. 15.2). Table 15.4 shows that, for strong values of the three synoptic-scale ascent mechanisms, we should add 20% for each mechanism to the climatological PoP (30%), resulting in a total PoP of 90% for Chicago. As we saw in Chapter 9, Chicago saw large amounts of ascent and heavy snow during this time period, substantiating our methodology. Finally, because Chicago is located in a region of mostly flat terrain, upslope and downslope flow were not considered.

In Chapter 9, we saw that Fort Smith is located underneath the convergent entrance region of a cyclonically curved jet and strong upper-tropospheric AVA, and in strong lower-tropospheric CAA (Fig. 15.2), mechanisms that are all associated with descent. As such, we subtract 20% for each descent mechanism from the climatological PoP (30%), resulting in a −30% PoP. Obviously, a negative PoP is not realistic, but the point is that the chance of precipitation at any location with three strong descent mechanisms is

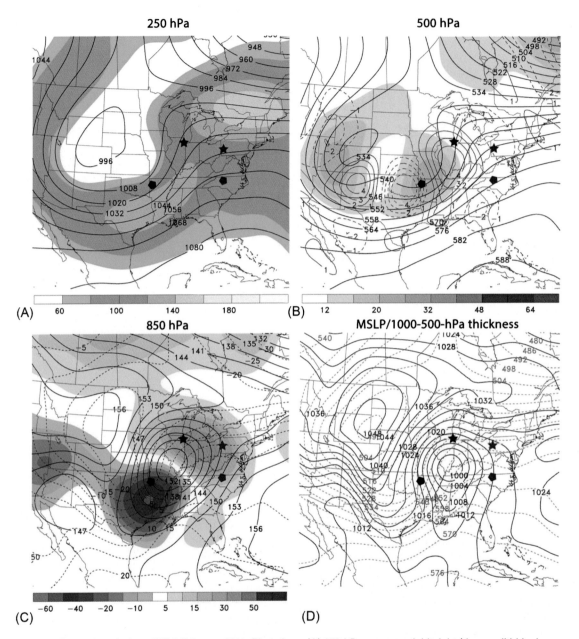

Fig. 15.2 Case example from 00Z 2 February 2011. Plotted are (A) 250-hPa geopotential height (dam, *solid black contours*) and wind speed (kt, *shaded*), (B) 500-hPa geopotential height (dam, *solid black contours*), geostrophic absolute vorticity ($\times 10^{-5}$ s^{-1}, *shaded*), and geostrophic absolute vorticity advection ($\times 10^{-9}$ s^{-2}, *blue contours, solid* for CVA, *dashed* for AVA), (C) 850-hPa geopotential height (dam, *solid black contours*), temperature (°C, *dashed black contours*), and geostrophic horizontal temperature advection ($\times 10^{-5}$ K s^{-1}, *shaded warm colors* for WAA, *cool colors* for CAA), and (D) mean sea-level pressure (hPa, *solid black contours*) and 1000–500-hPa thickness (dam, *dashed red contours*). As in Fig. 9.1, in each panel Chicago, Illinois and Fort Smith, Arkansas are marked with a *black star* and *pentagon*, respectively. Additionally, Pittsburgh, Pennsylvania and Greensboro, North Carolina are marked in each panel with a *purple star* and *pentagon*, respectively.

negligible. Because the surface winds (Fig. 15.2D) are generally from the north-northwest—a region of relatively flat terrain—upslope and downslope flow were not considered to be factors at Fort Smith.

Our third city, Pittsburgh, is located in a region of neutral jet divergence, weak upper-tropospheric CVA, and weak lower-tropospheric WAA (Fig. 15.2). In addition, the surface winds at Pittsburgh are generally easterly (Fig. 15.2D), which indicates some downsloping off of the Appalachians. Here, we add 20% to the climatological PoP (30%) for the combination of weak CVA and weak WAA, but subtract 15% for weak downsloping, resulting in a final PoP of 35%. An examination of METAR reports from 00 to 06Z on 2 February 2011 shows that Pittsburgh experienced mostly cloudy skies with occasional light rain and drizzle. This is consistent with our PoP calculation.

Lastly, we examine the conditions at Greensboro, located in the lee of the Appalachians. As at Pittsburgh, jet divergence and upper-tropospheric vorticity advection are neutral (Fig. 15.2). Weak WAA is present, and the surface winds blowing at an angle in toward the surface cyclone center are southeasterly but light, indicating weak upslope flow (Fig. 15.2D). As a result, we can add a total of 25% (10% for weak WAA and 15% for weak upslope flow) to our climatological PoP (30%), resulting in a PoP of 55% for Greensboro. METAR reports from 00 to 06Z on 2 February 2011 show overcast skies with light rain for 2 h and moderate rain for 1 h. In all, heavier and larger amounts of precipitation fell at Greensboro than Pittsburgh, substantiating our PoP analysis.

The PoP forecasting methodology discussed here is primarily intended to be a thought exercise for the novice forecaster, and should not replace more advanced or modern methods, including the use of ensemble NWP forecasts and blends. While simplistic and qualitative, it necessitates that the forecaster qualitatively assess all large-scale ascent and descent mechanisms prior to making his/her precipitation forecast. This technique is most effective in nonconvective midlatitude scenarios such as the winter storm discussed in this section. In the next section, we give a brief overview of how one should approach forecasting for heavy precipitation and flash flood events.

15.2.2 Ingredient-Based Approach

Although forecasting specific precipitation amounts is beyond the scope of this introductory text, it is important that the forecaster have an idea of the ingredients that are integral to causing heavy precipitation and flash flood events. To this point, we have focused primarily on ascent mechanisms, which are important, but do not tell the full story. In a scientific publication in 1996,[1] Dr. Chuck Doswell and coauthors outlined the three necessary ingredients for flash flooding: lift, moisture, and instability. This ingredient-based methodology can be applied to precipitation forecasts in any location or season, but is particularly useful in the warm season when instability and moisture tend to be more plentiful.

We have discussed all three ingredients in this text; ascent mechanisms in this chapter and in Chapter 9, and moisture and instability in the context of sounding analysis in Chapters 13 and 14. For example, a situation in which there are three large-scale mechanisms associated with ascent, a saturated troposphere with large dew points throughout the column (moisture), and a large amount of CAPE (instability, see Chapter 13) will result in more precipitation than would a situation with equivalent ascent, but smaller amounts of moisture and instability. Finally, it is useful to consider the duration of all three ingredients when making a precipitation forecast. The longer the duration that all three ingredients are present, the more precipitation that tends to result.

15.3 Questions and Practice Exercises

1. Calculate the forecast "max max" surface temperature for a summer afternoon at Regina, Saskatchewan (elevation of 577 m), where your favorite NWP model is predicting the following parameters:
   - 850-hPa temperature: 12°C.
   - 850-hPa geopotential height: 1570 m.
2. For the situation at Regina described in Question 1, the weather conditions listed below are predicted. Make the necessary adjustments to your "max max" calculation to come up with the best possible surface temperature forecast. Justify each adjustment that you make.
   - Partial cirrus cover, but no other clouds.
   - West wind of 15–20 kt.
3. You are attempting to make a minimum temperature forecast for Tampa, Florida on some night in September. After calculating your "max min" temperature, you assess that the following weather conditions will be present. Identify the impact of each condition on the forecast minimum temperature and explain your answers.
   - Overcast skies, primarily composed of low clouds.
   - West wind of 5–7 kt.
   - Occasional rain showers.

[1] Doswell III, C. A., H. E. Brooks, and R. A. Maddox, 1996: Flash flood forecasting: An ingredients-based methodology. *Wea. Forecasting*, **11**, 560–581, http://dx.doi.org/10.1175/1520-0434(1996)011,0560:FFFAIB.2.0.CO;2.

4. What are the three necessary ingredients for heavy precipitation and flash flooding?

5. You are attempting to make a PoP forecast for Toronto, Ontario. Below are the ascent mechanisms that you are predicting during your forecast period. Assuming the climatological PoP is 30%, what is your forecast PoP for this situation?
   - Weak jet streak divergence.
   - Strong CVA.
   - No temperature advection.
   - Minimal terrain effects.

6. Using the ingredient-based methodology for heavy precipitation, assess the flash flood potential of the Nashville, Tennessee sounding in Fig. 15.3. You may assume that upper-tropospheric vorticity advection is negligible, but that Nashville is located underneath the left exit quadrant of a straight jet streak. Explain your answer by discussing the appropriate physical mechanisms.

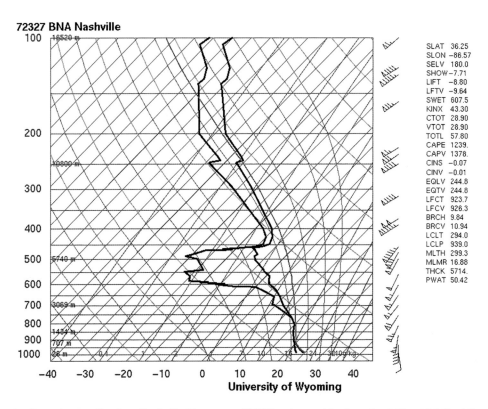

Fig. 15.3 Radiosonde sounding from Nashville, Tennessee (KBNA) to be used for Question 6. The *thick black line* on the right (left) side represents temperature (dew point). Reproduced with permission from the University of Wyoming Department of Atmospheric Science sounding archive, available online at: http://weather.uwyo.edu/upperair/sounding.html.

INDEX

Note: Page numbers followed by *f* indicate figures and *t* indicate tables.

Printed in the United States
By Bookmasters